Math Mammoth
Grade 4 Answer Keys

for the complete curriculum
(Light Blue Series)

Includes answer keys to:

- Worktext part A
- Worktext part B
- Tests
- Cumulative Reviews

By Maria Miller

Contents

Math Mammoth Grade 4-A
Answer Key

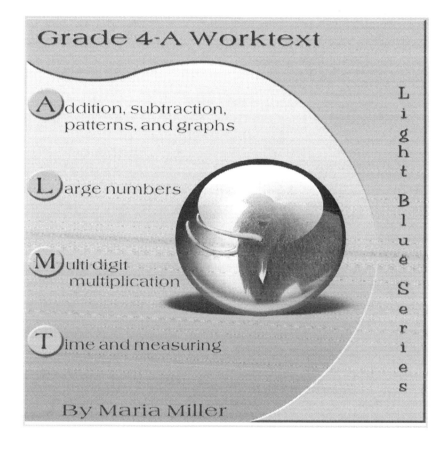

By Maria Miller

Math Mammoth Grade 4-A Answer Key
Contents

Chapter 1: Addition, Subtraction, Patterns, and Graphs

Addition Review, p. 11

1. a. 150, 157, 159   b. 190, 191, 199
   c. 110, 119, 120   d. 170, 175, 179

2. a. 400 + 80 + 7   b. 2,000 + 100 + 3
   c. 8,000 + 40 + 5   d. 600 + 50

3. a. It was 44. 56 + 90 + 44 = 190
   b. 70 + 80 = 150

4. a. 15, 65, 150, 1500   b. 13, 43, 130, 330
   c. 14, 24, 1400, 640

5. For example 50 + 80 = 130;
   500 + 800 = 1,300; 25 + 8 = 33

6. a. 87 + 34 + 44 = 165, 5 + 2 + 4 = 11,
   154 + 11 = 165
   b. 127 + 500 + 90 = 717, 4 + 3 + 9 = 16,
   717 + 16 = 733

7. Add one hundred then subtract one.
   a. 56 + 100 = 156; 156 − 1 = 155,
   b. 487 + 100 = 587, 587 − 1 = 586

8. a. 153, 79, 121   b. 89, 128, 111   c. 181, 101, 149

9.

Half the number	_10_	45	55	60	240	450	800	2,005
Number	20	90	110	120	480	900	1,600	4,010
Its double	_40_	180	220	240	960	1,800	3,200	8,020

10. a. $116   b. $105

11.

n	56	156	287	569	950	999
$n + 999$	1,055	1,155	1,286	1,568	1,949	1,998

12. a. 1,200, 1,800, 2,400, 3,000, 3,600, 4,200
    It reminds me of the multiplication table of 6.
    b. 1,800, 2,700, 3,600, 4,500, 5,400, 6,300
    It reminds me of the multiplication table of 9
    c. 175, 250, 325, 400, 475, 550, 625, 700

Adding in Columns, p. 14

1. a. 5,539   b. 9,058   c. 8,683

2. a. 8,325   b. 5,657

3. a. 672 miles   b. 261 miles

Subtraction Review, p. 15

1. a. 6, 56   b. 6, 76   c. 6, 60   d. 8, 800

2. a. 98, 80, 78   b. 196, 160, 155
   c. 495, 450, 444   d. 393, 330, 329

3. a. 4, 34, 40, 440   b. 6, 66, 60, 560
   c. 6, 66, 600, 360

4. Answers vary. Examples: 34 − 8 = 26, 140 − 80 = 60,
   240 − 80 = 160, 740 − 80 = 660.

5.

n	125	293	404	487	640	849
$n − 99$	26	194	305	388	541	750

6. a. 9, 5, 13   b. 18, 44, 48
   c. 27, 22, 46   d. 70, 50, 440
   e. 445, 944, 792   f. 418, 542, 492

7. a.

n	120	140	160	180	200	220	240	260	280
$n − 27$	93	113	133	153	173	193	213	233	253

   b. Each answer ends in 3, and each answer is 20 more
   than the previous answer.

8. a. 240, 200, 160, 120, 80, 40.
   It reminds me of the multiplication table of 4.
   b. 5,400, 4,800, 4,200, 3,600, 3,000, 2,400.
   It reminds me of the multiplication table of 6.
   c. 490, 420, 350, 280, 210, 140.
   It reminds me of the multiplication table of 7.

9. Game:
   a. 21 − 5 − 5 − 5 − 5 = 1
   b. 37 − 10 − 10 − 10 = 7
   c. 37 − 12 − 12 − 12 = 1 and
   50 − 7 − 7 − 7 − 7 − 7 − 7 − 7 = 1
   d. 30 − 9 − 9 − 9 = 3 and 20 − 8 − 8 = 4.

Subtract in Columns, p. 18

1. a. 173   b. 3,809   c. 568
   d. 344   e. 3,764   f. 5,326
   g. 217   h. 305   i. 5,580

2. a. 1,162   b. 4,925

3. First add 592, 87, 345 and 99; then subtract the sum
   from 5,200. The answer is 4,077

4. a. 1,530 miles. One round trip is 255 + 255 = 510 miles.
   Three round trips are 510 + 510 + 510 = 1,530 miles.

   b. 74 miles longer.

Puzzle Corner: 4:25

Patterns and Mental Math, p. 21

1. a.

n	9	18	27	36	45	54	63	72	81	90
$n + 29$	38	47	56	65	74	83	92	101	110	119

   b. The skip-counting pattern by 9's.
   c. Yes. There is a skip-counting pattern by 9's in the bottom row too, but it starts at 38.

2. a. *Hint: instead of subtracting 39, subtract 40, and add 1!*

n	660	600	540	480	420	360	300	240
$n - 39$	621	561	501	441	381	321	261	201

   b. It is a skip-counting pattern by 60's, backwards.
   c. Yes. It also has a skip-counting pattern going backwards by 60's.

3. a. 497, 470, 200, 467, 197   b. 598, 580, 400, 578, 398
   c. 993, 930, 300, 923, 293

4.

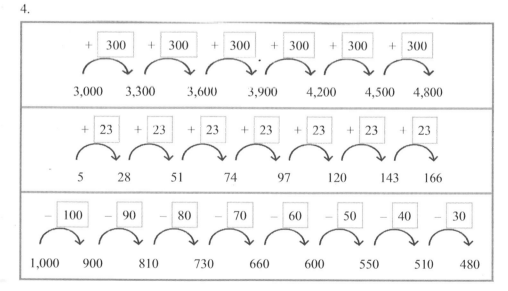

5. Subtract a thousand, then add one. To do 1,446 − 999, first subtract a thousand: 1,446 − 1,000 = 446.
   Then add one: 446 + 1 = 447.

6. a. $30. The second alarm clock cost $11 + $8 = $19, and $11 + $19 = $30.
   b. 50 days. June has 30 days and July has 31. There are 25 days with no rain in June and 25 days in July; a total of 50 days.
   c. 28 cm. The difference is 162 cm − 134 cm = 28 cm.
   d. Jack rode 100 km in all. 28 + 28 + (28 − 6) + (28 − 6) = 100 km
   e. 9 more boys. There are 45 − 18 = 27 boys, and 27 − 18 = 9 more boys than girls.

Patterns in Pascal's Triangle, p. 23

1.
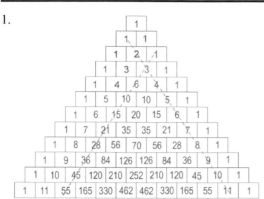

2. The row sums are:
1, 2, 4, 8, 16, 32, 64, 128, 256, 512, 1024, 2048.
These numbers double each time.

3. Yes - it is marked with a dashed line in the image.

4. a. 1, 3, 6, 10, 15, 21, 28, 36, 45, 55
   b. The diagonal is marked with a dashed line in the image.
   c. When you look at the differences of neighboring numbers, you get:

   2, 3, 4, 5, 6, 7, 8, 9, 10, and so on—which are the counting numbers.

Bar Models in Addition and Subtraction, p. 25

1.

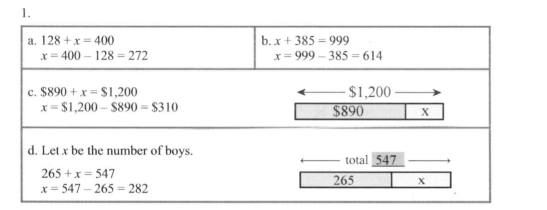

a. $128 + x = 400$ $x = 400 - 128 = 272$	b. $x + 385 = 999$ $x = 999 - 385 = 614$

c. $\$890 + x = \$1,200$
$x = \$1,200 - \$890 = \$310$

$\longleftarrow \$1,200 \longrightarrow$
$\$890$ | X

d. Let x be the number of boys.

$265 + x = 547$
$x = 547 - 265 = 282$

$\longleftarrow$ total 547 $\longrightarrow$
265 | x

2.

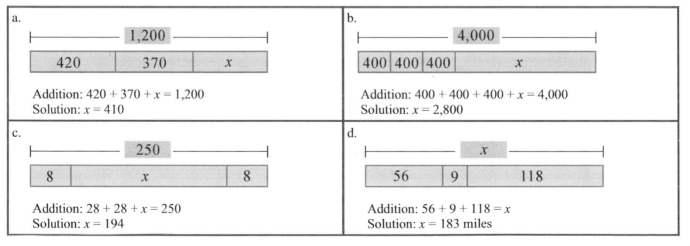

a.
1,200
420 | 370 | x
Addition: $420 + 370 + x = 1,200$
Solution: $x = 410$

b.
4,000
400 | 400 | 400 | x
Addition: $400 + 400 + 400 + x = 4,000$
Solution: $x = 2,800$

c.
250
8 | x | 8
Addition: $28 + 28 + x = 250$
Solution: $x = 194$

d.
x
56 | 9 | 118
Addition: $56 + 9 + 118 = x$
Solution: $x = 183$ miles

3. Answers vary. For example: A swimming pool costs $4,900. A family has saved $1,750 for it.
How much more do they still need to save? $x + 1,750 = 4,900$ $x = 4,900 - 1,750 = 3,150$

4. a. $x - 29 = 46$; $x = 29 + 46 = 75$
   b. Answers will vary. For example: $x - 255 = 99$ OR $x - 99 = 255$; $x = 255 + 99 = 354$

5. a. 24   b. 32   c. 99   d. 15   e. 72   f. 475

Bar Models in Addition and Subtraction, continued

6.

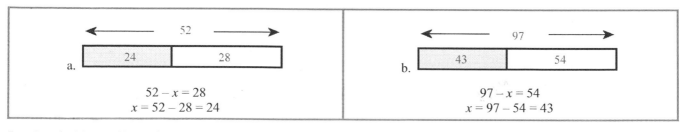

a.
$$52 - x = 28$$
$$x = 52 - 28 = 24$$

b.
$$97 - x = 54$$
$$x = 97 - 54 = 43$$

7. a. 8   b. 21   c. 134   d. 18   e. 28   f. 557

8. a. $15 + x = 22; $x = 7 b. $x - 24 = 125$; $x = 149$
   c. $120 - x = 89$; $x = 31$ d. $x - 67 = 150$; $x = 217$

9. a. $x + 43 = 450$; Subtract to solve x. $x = 450 - 43 = 407$
   b. $250 - x = 78; Subtract to solve x. $x = 250 - 78 = 172
   c. $200 - $54 - $78 = x$; Subtract to solve x. $x = 68
   d. $x - $23 - $29 = 125; Add to solve x. $x = $125 + $29 + $23 = 177

Puzzle Corner: a. 60   b. 220   c. $x = 31$ d. $x = 37$

In (a), we do know the TOTAL (subtraction always starts with the total), and one of the PARTS is missing. To find the missing part, subtract the other parts from the total. So, to solve $200 - 45 - \_\_\_ - 70 = 25$, subtract the other "parts" (45, 70, and 25) from 200.

In (b), the total is missing: $\_\_\_ - 5 - 55 - 120 = 40$. We can find it by adding all the parts (the 5, 55, 120, and 40).

In (c) and (d), we have missing addend problems that are solved by subtracting all the "parts" from the total.

Order of Operations, p. 29

1. a. 440, 500, 500, 440   b. 350, 350, 150, 250

2. a. 16, 11, 4   b. 13, 22, 5   c. 19, 14, 49

3. $90 - 2 \times 20 = 50$. A 50-cm piece is left.

4. a. $5 \times 10 - 7 = 43$;   b. $(100 - 20) + 10 = 90$;
   c. $5 \times (10 - 7) = 15$; d. $100 - (20 + 10) = 70$

5. a. and c.

6. Answers will vary - for example:
   a. Anne buys a shirt for $10, a box of pens for $2.10 and a jacket for $45. What is the total cost?
      $10 + $2.10 + 45 = 57.10
   b. Tim bought four ice cream cones for $1.20 each. What was the total cost? $4 \times $1.20 = 4.80
   c. Tim bought four ice cream cones for $1.20 each, and paid with $10. What was Tim's change?
      The change was $10 - 4 \times $1.20 = 5.20.

7. a. $4 \times 1 + 8 = 12$;
      $50 - 5 \times 10 = 0$ or $50 \div 5 - 10 = 0$
   b. $2 + 10 + 1 \times 2 = 14$; $100 - (15 + 17) \times 1 = 68$
   c. $3 \times 3 - 3 = 6$; $(2 + 5) \times 2 = 14$

There may be other correct number sentences for the problems 8 - 10.

8. $100 \text{ kg} - 4 \times 5 \text{ kg} = 80 \text{ kg}$
   or $100 \text{ kg} - 5 \text{ kg} - 5 \text{ kg} - 5 \text{ kg} - 5 \text{ kg} = 80 \text{ kg}$
   or $100 \text{ kg} - (5 \text{ kg} + 5 \text{ kg} + 5 \text{ kg} + 5 \text{ kg}) = 80 \text{ kg}$

9. $5 \times $2 + 2 \times $3 = 16 *or* $2 \times $3 + 5 \times $2 = 16
   or $2 + $2 + $2 + $2 + $2 + $3 + $3 = 16

10. $20 - $7 - $5 = 8 *or* $20 - ($7 + $5) = 8

Making Bar Graphs, p. 31

1. a.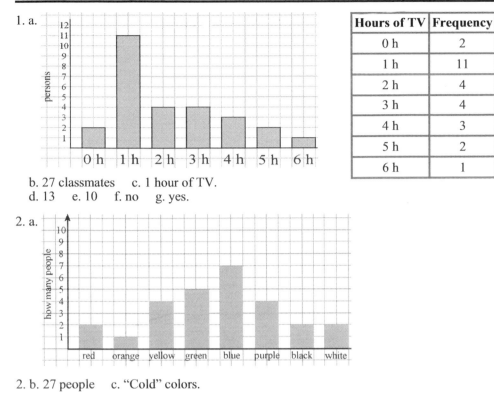

Hours of TV	Frequency
0 h	2
1 h	11
2 h	4
3 h	4
4 h	3
5 h	2
6 h	1

b. 27 classmates   c. 1 hour of TV.
d. 13   e. 10   f. no   g. yes.

2. a.

2. b. 27 people   c. "Cold" colors.

3. a.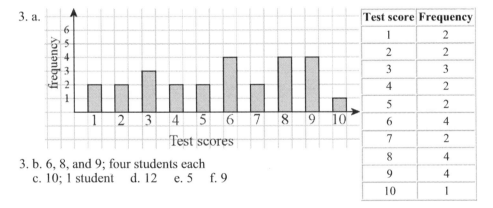

Test score	Frequency
1	2
2	2
3	3
4	2
5	2
6	4
7	2
8	4
9	4
10	1

3. b. 6, 8, and 9; four students each
c. 10; 1 student   d. 12   e. 5   f. 9

Line Graphs, p. 33

1. a. $60   b. $140   c. June   d. $15   e. $70

2. a. Day 1: 500 grams; Day 2: 525 grams;
Day 3: 550 grams; Day 4: 575 grams
b. Day 5.
c. Day 8.

3. a. The price lowers in the summer and is higher in
the winter. That is because in the summer there is an
abundance of strawberries; all stores and markets are
selling strawberries. Nobody can keep the price high
because if they did, people would go elsewhere to buy.

3. b. The highest price was in December, $3.60 per pound,
and the lowest price was in July, $1.63 per pound.
The difference is $1.97.

c. In August, $3.64. In November, $6.38.

Line Graphs, continued

4. a.

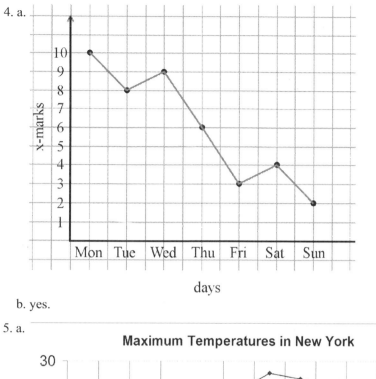

days

b. yes.

5. a.

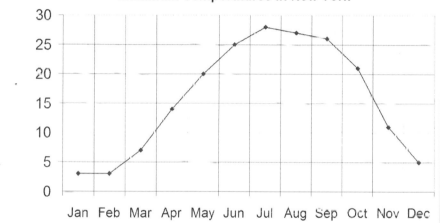

b. January, February, and December   c. June, July, August, and September   d. 25 degrees

Rounding, p. 36

1. a. 290   b. 320   c. 280   d. 290   e. 320   f. 300   g. 300   h. 210

2. a. 530   b. 30   c. 180   d. 200   e. 710   f. 390
   g. 440   h. 5,970   i. 9,570   j. 4,060   k. 2,280   l. 4,000

3. a. 3,500   b. 3,700   c. 3,900   d. 3,500   e. 4,000   f. 3,400

4. a. 500   b. 100   c. 800   d. 200   e. 700   f. 400
   g. 2,900   h. 6,000   i. 7,500   j. 3,000   k. 3,000   l. 4,000

5. a. 4,000   b. 7,000   c. 5,000   d. 7,000   e. 3,000   f. 4,000

6. a. 1,000   b. 0   c. 1,000   d. 4,000   e. 6,000   f. 3,000
   g. 3,000   h. 6,000   i. 9,000   j. 10,000   k. 3,000   l. 1,000

7.

n	55	2,602	9,829	495	709	5,328
rounded to nearest 10	60	2,600	9,830	500	710	5,330
rounded to nearest 100	100	2,600	9,800	500	700	5,300
rounded to nearest 1000	0	3,000	10,000	0	1,000	5,000

Estimating, p. 39

1. a. Estimation: $1,000 + 200 + 4,800 = 6,000$ Exact: 5,990
   b. Estimation: $300 + 400 + 600 = 1,300$ Exact: 1,312
   c. Estimation: $1,000 - 400 - 100 = 500$ Exact: 552
   d. Estimation: $3,500 - 1,500 - 200 = 1,800$ Exact: 1,741

2. About $4 \times 50 = 200$ passengers.

3. About $150 + $160 + $180 + $130 + $130 = $750

4. a. Apartment 1: about $290 + $290 + $290 = $870
      Apartment 2: about $330 + $330 + $330 = $990.
   b. They would save approximately: $990 − $870 = $120.

5. a. $340 + 360 + 320 + 320 = 1,340$
   b. $300 + 290 + 290 + 260 = 1,140$

1. a. 25¢   b. 178¢   c. 1560¢   d. $0.20   e. $1.54   f. $8.59

2. a. $2.20 + $12 + $1.50 = $15.70.   b. $20 − $2.20 − $12 − $1.50 = $4.30

3.

Item cost	Money given	Change needed	$50 bill	$20 bill	$5 bill	$1 bill
a. $56	$70	$14			2	4
b. $78	$100	$22		1		2
c. $129	$200	$71	1	1		1

4.

Item cost	Money given	Change needed	$5 bill	$1 bill	25¢	10¢	5¢	1¢
a. $2.56	$5	$2.44		2	1	1	1	4
b. $7.08	$10	$2.92		2	3	1	1	2
c. $3.37	$10	$6.63	1	1	2	1		3

5. a. **ii.** x = $12   b. **iii.** x = $108   c. **i.** $12

6. b. x − $250 = $170; x = $420

   c. x − $45 = $15; x = $60

   d. $12 − x = $3.56; x = $8.44

   e. x − $12 − $9 = $29; x = $50

   f. $65 − $12 − $12 − $7 = x; OR $65 − 2 × $12 − $7 = x; x = $34

   g. $20 − x − x = $12.40 OR $20 − 2x = $12.40; x = $3.80

   h. $50 − x − x − x = $17 OR $50 − 3x = $17; x = $11.

7. $999

8. $19

9. a. $0.75   b. $0.24   c. $477/month   d. $35

10. a. $54.99 − x = $47.99; x = $7

    b. x − $35 = $94; x = $129

    ←——— original price $129 ——→

$94	$35

Calculate and Estimate Money Amounts, p. 44

1. a. $6.30   b. $10.00   c. $5.60   d. $0.30   e. $0.70   f. $5.00

2. a. $3   b. $98   c. $3   d. $1,680   e. $47   f. $126

3. a. $50   b. $10   c. $70   d. $6,290   e. $40   f. $170

4.

n	$29.78	$5.09	$59.95	$2.33	$0.54
rounded to nearest ten cents	$29.80	$5.10	$60.00	$2.30	$0.50
rounded to nearest dollar	$30.00	$5.00	$60.00	$2.00	$1.00

5. a. $2 + $6 + $5 + $13 = $26   b. $120 + $200 + $260 + $340 = $920

6. Answers may vary according to the rounding.
   a. about $12 (rounding to the nearest dollar) or about $11.90 (rounding to the nearest ten cents)
   b. six gallons
   c. about 5 × $2 + 2 × $5 = $20
   d. five ice cream cones

7. a. $23.30   b. $369.50   c. $201.01   d. $30.75

8. a. A 1-Day Adult's ticket costs $6 more than a 1-Day Children's ticket.
   b. A 2-Day Adult's ticket costs $12 more than a 2-Day Children's ticket.
   c. For a 4-Day Adult's ticket, the discount is $2.48.
   d. For a 4-Day Children's ticket, the discount is $3.32.

9. a. $929.58 is the total for 3 days using discount tickets.
   b. Yes, the total would be $978.40 for 4-day discount tickets.

Review, p. 47

1.

a. 81 − 72 = 9 665 − 99 = 566	b. 45 + 65 =110 196 + 99 = 295	c. 160 + 280 = 440 54 − 28 = 26

2. $x + 38 = 230$; $x = 192$ ←—— total 230 ——→

192	38

3. $x + 587 = 1,394$; $x = 1,394 − 587 = 807$

4. a. 30, 70   b. 100, 29   c. 82, 76

5. ($13 − $2) × 3 = $33.

6. (10 × 4) + (20 × 2) = 80 feet altogether.

7. $25 + $14 + $3 = $42

8. $15.20 + $34.60 + $70.20 = $120

9. $48.90 + ($48.90 + $25) = $122.80

Chapter 2: Large Numbers and Place Value

Thousands, p. 51

1. b. $4,000 + 900 + 30 + 5$ c. $4,000 + 0 + 30 + 9$ d. $3,000 + 0 + 0 + 2$
   e. $2,000 + 0 + 90 + 0$ f. $9,000 + 400 + 0 + 5$

2. a. 4,593   b. 2,090   c. 3,200   d. 8,005   e. 4,600   f. 4,080
   g. 7,203   h. 1,405   i. 7,050   j. 4,005   k. 4,069   l. 3,809

3. b. five hundred   c. five thousand   d. fifty

4. b. nine thousand   c. forty   d. eighty   e. two hundred   f. two   g. twenty   h. five

5. a. 8,542   b. 2,458

6. Difference is: $961 - 169 = 792$.

7.

n	2,508	342	4,009	59	6,980	8,299
$n + 10$	2,518	352	4,019	69	6,990	8,309
$n + 100$	2,608	442	4,109	159	7,080	8,399
$n + 1000$	3,508	1,342	5,009	1,059	7,980	9,299

8. a. 6   b. 400   c. 8   d. 8,000   e. 5,000   f. 600

9. 7,889

Puzzle Corner: b. 29 and 92; difference: 63   c. 45 and 54; difference: 9   d. 38 and 83; difference: 45
e. You can find all those in the multiplication table of 9.
f. 0 and 4, 1 and 5, 2 and 6, 3 and 7, 4 and 8, 5 and 9
g. 0 and 3, 1 and 4, 2 and 5, 3 and 6, 4 and 7, 5 and 8, or 6 and 9
h. 0 and 9

At the Edge of Whole Thousands, p. 54

1. a. $991 + 9 = 1,000$ b. $960 + 40 = 1,000$ c. $942 + 58 = 1,000$
   d. $924 + 76 = 1,000$ e. $933 + 67 = 1,000$ f. $979 + 21 = 1,000$

2. a. $999 + 1 = \underline{1,000}$; $992 + \underline{8} = \underline{1,000}$ b. $980 + \underline{20} = \underline{1,000}$; $985 + \underline{15} = \underline{1,000}$ c. $930 + \underline{70} = \underline{1,000}$; $937 + \underline{63} = \underline{1,000}$

3. a. $1,920 + 80 = 2,000$; $1,999 + 1 = 2,000$; $2,998 + 2 = 3,000$
   b. $1,990 + 10 = 2,000$; $7,940 + 60 = 8,000$; $5,970 + 30 = 6,000$
   c. $6,950 + 50 = 7,000$; $4,900 + 100 = 5,000$; $3,995 + 5 = 4,000$

4. a. 1,999; 1,996; 1,993   b. 4,997; 3,990; 6,980   c. 5,950; 8,970; 9,900

5.

Number	Rounded number	Rounding error
4,993	5,000	7
7,890	8,000	110
9,880	10,000	120

Number	Rounded number	Rounding error
8,029	8,000	29
5,113	5,000	113
2,810	3,000	190

6. a. 1,900; 1,850; 1,750   b. 4,800; 4,770; 4,720   c. 8,500; 8,420; 8,320

7. a. $3,000   b. It is $3,000 - $2,992 = $8 short.

8. The error is 68. The first addend, 1982, is 18 less than 2,000, and the second addend, 3,950 is 50 less than 4,000.
This creates a total rounding error of $18 + 50 = 68$.

More Thousands, p. 56

1. a. 164,000, 164 thousand   b. 92,000, 92 thousand   c. 309,000, 309 thousand
   d. 34,000, 34 thousand   e. 780,000, 780 thousand

2. b. 92,908, 92 thousand 908   c. 329,033, 329 thousand 033   d. 14,004, 14 thousand 004
   e. 550,053, 550 thousand 053 f. 72,001, 72 thousand 001
   g. 800,004, 800 thousand 004 h. 30,036, 30 thousand 036

3. a. four hundred fifty-six thousand ninety-eight
   b. nine hundred fifty thousand fifty
   c. twenty-three thousand ninety
   d. five hundred sixty thousand eight
   e. seventy-eight thousand three hundred four
   f. two hundred sixty-six thousand eight hundred ninety-four
   g. one million
   h. three hundred six thousand seven hundred

4. a. 35,000   b. 201,000   c. 430,000   d. 750,000   e. 1,000,000   f. 770,000

5. a. 40,000   b. 721,000   c. 450,000   d. 630,000   e. 240,000
   f. 800,000   g. 25,000   h. 194,000   i. 323,000   j. 499,000

6. The blue dots on the number line mark the numbers 502,000, 511,000, 524,000, and 538,000.

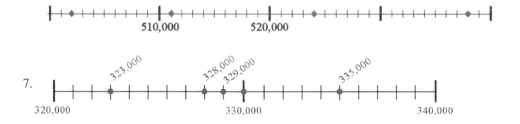

7.

Practicing with Thousands, p. 58

1. a. 49 thousands 0 hundreds 1 ten 5 ones   b. 206 thousands 0 hundreds 9 tens 0 ones
   c. 107 thousands 8 hundreds 0 tens 2 ones   d. 88 thousands 0 hundreds 3 tens 0 ones
   e. 790 thousands 3 hundreds 0 tens 2 ones   f. 903 thousands 0 hundreds 0 tens 0 ones
   g. 250 thousands 0 hundreds 6 tens 7 ones   h. 300 thousands 0 hundreds 7 tens 0 ones

2. a. 20,704   b. 204,080   c. 101,600   d. 540,004   e. 230,370
   f. 9,607   g. 873,050   h. 40,400   i. 59,065

3. a. 25,347   b. 700,624   c. 61,808   d. 53,060   e. 42,087
   f. 1,000,000 g. 290,040   h. 27,905   i. 504,008

4.

a.	b.	c.	d.
45,000	134,000	800,000	400,000
45,500	134,200	750,000	390,000
46,000	134,400	700,000	380,000
46,500	134,600	650,000	370,000
47,000	134,800	600,000	360,000
47,500	135,000	550,000	350,000
48,000	135,200	500,000	340,000
48,500	135,400	450,000	330,000
49,000	135,600	400,000	320,000

5. a. 30,050   b. 254,305   c. 133,250   d. 77,004   e. 60,002
   f. 120,063   g. 15,020   h. 24,006   i. 30,390   j. 86,471

Place Value with Thousands, p. 60

1. a.

hth	tth	th	h	t	o
	8	7, 0	1	5
	8	0, 0	0	0
		7, 0	0	0
		0	0	0
				1	0
					5

b.

hth	tth	th	h	t	o
4	0	3, 2	8	0
4	0	0, 0	0	0
	0	0, 0	0	0
		3, 0	0	0
		2	0	0
			8	0
				0

c.

hth	tth	th	h	t	o
6	9	2, 0	0	4
6	0	0, 0	0	0
	9	0, 0	0	0
		2, 0	0	0
		0	0	0
			0	0
				4

d.

hth	tth	th	h	t	o
7	0	0, 2	0	4
7	0	0, 0	0	0
	0	0, 0	0	0
		0, 0	0	0
		2	0	0
			0	0
				4

2. a. 80,000 + 7,000 + 10 + 5   b. 400,000 + 3,000 + 200 + 80

3. a. 30,000 + 2,000 + 400 + 90 + 3   b. 100,000 + 70,000 + 2,000 + 300 + 90 + 2   c. 20,000 + 5,000 + 600
   d. 100,000 + 9,000 + 20   e. 900,000 + 700 + 1

4. a. 26,000   b. 5,000   c. 20   d. 7,000   e. 70,000   f. 7,000

5. b. five   c. five hundred   d. fifty thousand

6. a. ten thousand   b. three hundred thousand   c. eighty thousand   d. eight thousand
   e. six hundred   f. twenty   g. ten thousand   h. nine hundred thousand

7. a. 500,087   b. 20,480   c. 907,700   d. 50,360

Puzzle corner. 630,001

Comparing with Thousands, p. 62

1. a. <   b. <   c. <   d. >   e. <   f. <   g. <   h. >   i. <

2. a. 8,039 < 18,309 < 81,390 < 818,039   b. 5,020 < 52,000 < 250,000 < 520,000

3. a. 54,000   b. 8,708   c. 11,101   d. 144,000   e. 5,606   f. 8,909

4. a. >   b. <   c. >   d. >   e. >   f. >   g. <   h. >

5. a.

| 15,000 | 15,100 | 15,200 | 15,300 | 15,400 | 15,500 | 15,600 | 15,700 | 15,800 | 15,900 | 16,000 | 16,100 |

   b.

| 34,500 | 34,600 | 34,700 | 34,800 | 34,900 | 35,000 | 35,100 | 35,200 | 35,300 | 35,400 | 35,500 | 35,600 |

6. 67,030 < 67,049 < 67,250 < 67,370 < 67,510 < 67,703 < 67,780 < 67,940

7. a. 398,039   b. 290,290   c. 606,660   d. 110,293   e. 301,481   f. 390,200

8. a. 500 < 1,459 < 1,500 < 5,406 < 5,505 < 5,600
   b. 7,800 < 8,708 < 77,988 < 78,707 < 78,777 < 87,600

9. a. x = 200,000   b. x = 70,000   c. x = 5,000

Comparing with Thousands, continued

10. a.

n	600	1,200	1,800	2,400	3,000	3,600	4,200
$n + 500$	_1100_	1,700	2,300	2,900	3,500	4,100	4,700

    b. It is a skip-counting pattern by 600's, starting at 1,100. Just as the top numbers go by 600s, so do the bottom numbers. That is because when we add 500 from the top numbers, we are not changing the difference between each two numbers.

11. a.

n	52,000	55,000	58,000	61,000	64,000	67,000
$n - 5,000$	_47,000_	50,000	53,000	56,000	59,000	62,000

    b. The pattern is the number increases by 3,000 each time.

Adding and Subtracting Big Numbers, p. 65

1. a. 945,601   b. 436,929   c. 818,736   d. 149,259   e. 905,468   f. 524,160

2.

a.	b.	c.
29,100	906,500	610,400
29,300	916,600	610,000
29,500	926,700	609,600
29,700	936,800	609,200
29,900	946,900	608,800
30,100	957,000	608,400
30,300	967,100	608,000
30,500	977,200	607,600
30,700	987,300	607,200
30,900	997,400	606,800

3. a. 85,581   b. 119,976   c. 668,700   d. 66,697   e. 41,893   f. 85,055   g. 426,600   h. 376,935   i. 381,656

4.

a.

419,000 + 1,000 — 20,000 + 400,000
500 + 36,000 — 180,000 − 2,000
189,000 + 1,000 — 150,000 + 40,000
40,500 + 500 — 36,000 + 5,000
177,300 + 700 — 36,100 + 400

b.

500,000 − 3,000 — 97,000 + 400,000
189,000 − 80,000 — 100,000 + 9,000
40,600 − 500 — 20,000 + 20,100
250,000 − 40,000 — 140,000 + 70,000
77,700 − 7,000 — 100,000 − 29,300

5. a. 302,889   b. 641,571   c. 26,712   d. 876,255

6.

n	13,000	78,000	154,000	500,000	640,500
$n + 1,000$	14,000	79,000	155,000	501,000	641,500
$n + 10,000$	23,000	88,000	164,000	510,000	650,500
$n + 100,000$	113,000	178,000	254,000	600,000	740,500

7. a. 427,443   b. 27,854   c. 25,301   d. 983,715

8. a. >   b. =   c. >   d. =   e. <   f. =

1.

a. 45<u>2</u>,550 ≈ 450,000	b. 8<u>6</u>,256 ≈ 86,000	c. 77,<u>5</u>79 ≈ 77,600
d. 24<u>5</u>,250 ≈ 245,000	e. <u>8</u>94,077 ≈ 900,000	f. 385,<u>7</u>06 ≈ 386,000
g. <u>6</u>15,493 ≈ 600,000	h. <u>5</u>27,009 ≈ 500,000	i. <u>2</u>52,000 ≈ 300,000
j. <u>2</u>6,566 ≈ 30,000	k. 9<u>4</u>4,032 ≈ 940,000	l. 33<u>5</u>,700 ≈ 336,000
m. 48,4<u>2</u>1 ≈ 48,420	n. 8,<u>5</u>55 ≈ 8,600	o. 4<u>0</u>9,239 ≈ 410,000

2.

a. 10,<u>9</u>65 ≈ 11,000	b. 89,<u>5</u>06 ≈ 90,000	c. 79<u>7</u>,329 ≈ 800,000
d. 299,<u>8</u>50 ≈ 300,000	e. 254,99<u>7</u> ≈ 255,000	f. 599,9<u>7</u>2 ≈ 600,000

3.

a. 233,<u>5</u>64 ≈ 233,600	b. 75<u>2</u>,493 ≈ 752,000	c. 1<u>9</u>2,392 ≈ 190,000
d. 8<u>9</u>5,080 ≈ 900,000	e. <u>8</u>55,429 ≈ 900,000	f. 39<u>9</u>,477 ≈ 399,000

4.

number	274,302	596,253	709,932	899,430
to the nearest 1,000	274,000	596,000	710,000	899,000
to the nearest 10,000	270,000	600,000	710,000	900,000
to the nearest 100,000	300,000	600,000	700,000	900,000

5. a. about 1,800 days.   b. about 3,300 days.   c. about 3,700 days.   d. about 7,300 days.
e. In 40 years, you have likely lived 14,610 days, or about 14,600 days.
f. Answers vary. For example, if a mother is 36 years old, subtract four times 365 from 14,610.
   And since 365 + 365 = 730, we can subtract two times 730, or 1,460:
   14,610 − 1,460 = 13,150. This is about 13,200 days.

6. a. 235 ≈ 0   b. 18,299 ≈ 20,000   c. 1,392 ≈ 0

7. a. 865 ≈ 1,000   b. 182 ≈ 0   c. 5,633 ≈ 6,000

8.

a. 56,250 ≈ 60,000	b. 5,392 ≈ 10,000	c. 2,938 ≈ 0
d. 708,344 ≈ 710,000	e. 599 ≈ 0	f. 44,800 ≈ 40,000

Rounding and Estimating Large Numbers, continued

9.

a. That means about <u>236,000</u> people in Purpletown, and about <u>187,000</u> people in Bluetown. The two towns have approximately <u>423,000</u> people in all. There are about <u>49,000</u> more people in Purpletown than in Bluetown.	
b. There were about <u>3,500</u> live births in total in those two. Seagull hospital had about <u>1,300</u> more live births than Sunshine hospital.	
c. Rounding to the nearest thousand makes the problem simple. That means about $48,000. So he earns about <u>$4,000</u> monthly.	d. The simplest way is to round the annual mileage to the nearest 10,000, That is about <u>60,000</u> miles. This means he drives about <u>5,000 miles</u> each month.

10. a. The trip around the equator is about <u>25,000</u> miles.
An adult human's blood vessels go for about <u>93,000</u> miles.
So, how many whole loops around the earth would those blood vessels go?   <u>3</u> loops

b. 239,000

c. 3 adults would be needed. With two adults, we get 93 thousand + 93 thousand = 186 thousand, which is not enough. With three adults we get 186 thousand + 93 thousand = 279 thousand, which is more than 239,000.

d. Their white blood cell count is <u>high.</u> Their platelet count is <u>low.</u>

e. Answers vary. For example, you could answer you have about 99,900. Or, you could answer you still have about 100,000, because subtracting, you get 100,000 − 100 = 99,900, but that number is still rounded to 100,000 when rounding to the nearest thousand.

Should you worry? No, because new hair grows in to replace them. It would take nearly two years for half of your hair to fall out.

Multiples of 10, 100, and 1000, p. 73

1.

a. $11 \times 10 = 110$ $29 \times 10 = 290$	b. $50 \times 10 = 500$ $80 \times 10 = 800$	c. $200 \times 10 = 2,000$ $1000 \times 10 = 10,000$

2.

a. $7 \times 100 = 700$ $9 \times 100 = 900$	b. $10 \times 100 = 1,000$ $13 \times 100 = 1,300$	c. $20 \times 100 = 2,000$ $22 \times 100 = 2,200$

3.

a. $311 \times 100 = 31,100$ $70 \times 100 = 7,000$ $120 \times 100 = 12,000$	b. $10 \times 19 = 190$ $999 \times 10 = 9,990$ $10 \times 4,500 = 45,000$	c. $60 \times 1,000 = 60,000$ $493 \times 1,000 = 493,000$ $1,000 \times 500 = 500,000$

4.

a. 49 thousands 49,000 49 hundreds 4,900 49 tens 490	b. 20 tens 200 20 hundreds 2,000 20 thousands 20,000	c. 37 tens 370 37 hundreds 3,700 37 thousands 37,000

5.

a. 10 tens 100 100 tens 1,000	b. 10 hundreds 1,000 100 hundreds 10,000	c. 100 thousands 100,000 1,000 thousands 1,000,000

6. a. $4,000   b. $500   c. $20,000

7. a. 1492   b. $2,500; $1,200   c. 1900; 1960

8.

a. $67 \times 10 = 670$ $18 \times 100 = 1,800$ $20 \times 10 = 200$	b. $112 \times 100 = 11,200$ $80 \times 1,000 = 80,000$ $390 \times 10 = 3,900$	c. $100 \times 44 = 4,400$ $10 \times 90 = 900$ $1,000 \times 60 = 60,000$

9.

a. $120 \div 10 = 12$ $600 \div 10 = 60$ $1,300 \div 10 = 130$	b. $700 \div 100 = 7$ $5,600 \div 100 = 56$ $65,000 \div 100 = 650$	c. $12,000 \div 1000 = 12$ $689,000 \div 1000 = 689$ $400,000 \div 1000 = 400$

10.

a.

90	$= 9 \times 10$
100	$= 10 \times 10$
110	$= 11 \times 10$
280	$= 28 \times 10$
1000	$= 100 \times 10$
4,560	$= 456 \times 10$

b.

900	$= 9 \times 100$
1000	$= 10 \times 100$
1100	$= 11 \times 100$
4,000	$= 40 \times 100$
5,900	$= 59 \times 100$
10,000	$= 100 \times 100$

c.

10,000	$= 10 \times 1000$
15,000	$= 15 \times 1000$
18,000	$= 18 \times 1000$
50,000	$= 50 \times 1000$
160,000	$= 160 \times 1000$
520,000	$= 520 \times 1000$

11. Notice something special about these multiples: ALL multiples of 10 end in 0
ALL multiples of 100 end in 00. ALL multiples of 1000 end in 000.

12. Answers vary.

Multiples of ten	What number times 10?		Multiples of hundred	What number times 100?
540	$= 54 \times 10$		48,200	$= 482 \times 100$
9,870	$= 987 \times 10$		63,600	$= 636 \times 100$
42,090	$= 4209 \times 10$		849,000	$= 8490 \times 100$

13.

a. $1,000 \div 100 = 10$ $2,100 \div 100 = 21$ $99,900 \div 100 = 999$	b. $90 \div 10 = 9$ $7,000 \div 10 = 700$ $34,800 \div 10 = 3,480$	c. $2,000 \div 1,000 = 2$ $30,000 \div 1,000 = 30$ $342,000 \div 1,000 = 342$

Puzzle corner:
100,000   200,000   300,000

Mixed Review, p. 76

1. $23.50 + $19.90 + $6.60 = x$
   $x = 50

2. a. $56.25   b. $283.01

3. ($15 − $3) × 4

4. 5,209 < 25,539 < 25,925 < 525,009

5. a. $x = 47$ b. $x = 1,266$ c. $x = 633$

6. a. $158 − $38 = x$; Grandma gave him $120.
   b. $x − 3 × $4 = 28; 3 × $4 + $28 = $40.
   c. $60 − 2 × $11 = x$. He had $38 left.
   d. 3 × $0.60 + 1 × $0.80 = x$. His total was $2.60.
      He received $7.40 change.

7. a. 60,000 + 8,000 + 50 + 6
   b. 800,000 + 10,000 + 5,000 + 200 + 20 + 4

8. a. $1.00 b. $8.00 c. $35.00 d. $166.00
   e. $95.00 f. $99.00 g. $100.00 h. $101.00

9. a. $20.00 + $15.00 + $25.00 = $60.00 total.
   b. 4,000 × $1.00 = $4,000; 1,000 × $1.00 = $1,000.
      $4,000 + $1,000 = $5,000 total.

Review, p. 78

1. a. 13,094   b. 306,050   c. 1,000,000

2. a. 785,300   b. 70,008

3. a. three thousand   b. thirty   c. 300 thousand   d. 30 thousand

4.

n	78,974	5,367	2,558	407,409	299,603
rounded to nearest 1,000	79,000	5,000	3,000	407,000	300,000
rounded to nearest 10,000	80,000	10,000	0	410,000	300,000

5. Estimate: 5,100 − 2,800 − 700 = 1,600   Exact: 1,556

6. a. 500   b. 700,000   c. 600,000

7. a. 5,406 < 5,604   b. 49,530 < 49,553   c. 605,748 > 60,584

8. 95,695 < 145,900 < 495,644 < 496,455 < 590,554 < 5,905,544

9. a. 392,054   b. 444,869

10. 500 × $100 = $50,000

11. In ten months: 10 × $2,560 = $25,600. In two months: $2,560 + $2,560 = $5,120
    In 12 months: $25,600 + $5,120 = $30,720

Chapter 3: Multiplication

Understanding Multiplication, p. 84

1. a. $2 + 2 + 2 + 2 = 4 \times 2 = 8$; $20 + 20 + 20 + 20 = 4 \times 20 = 80$
   b. $8 + 8 + 8 = 3 \times 8 = 24$; $80 + 80 + 80 = 3 \times 80 = 240$;
   c. $500 + 500 + 500 + 500 = 4 \times 500 = 2,000$ $120 + 120 + 120 = 3 \times 120 = 360$

2. a. $6 \times 3 = 18$; $3 \times 6 = 18$ b. $5 \times 1 = 5$; $1 \times 5 = 5$

3. a. $16, 0$ b. $15, 10$ c. $16, 8$ d. $30, 27$

4. a. $48, 0, 16$ b. $300; 6,000; 12,000$ c. $4,000; 1,000; 633$ d. $68, 63, 200$

5. a. $6, 0$ b. $200, 250$ c. $3, 81$

6. a. factors; product   b. $4 \times 8 = 32$ c. The product is 0.   d. 5, because $2 \times 6 \times \underline{5} = 60$

7. a. $3 \times 12 + 5 = 41$ eggs in all.   b. Jack had: $6 \times 10 - 3 = 57$ left.
   c. $4 \times 10 + 3 \times 6 = 58$ crayons in all.   d. Ernest's change was: $\$50 - 3 \times \$11 = \$17$.
   e. $5 \times 3 + 7 \times 2 = 29$ wheels

8.

a. $7 \times 10 = ?$ $? = 70$	b. $? \times 4 = 24$ $? = 6$
c. $? \times 2 = 18$ $? = 9$	d. $y \times 4 = 36$ $y = 9$
e. $y \times 10 = 150$ $y - 15$	f. $y \times 12 = 60$ $y = 5$
g. $4 \times 20 = y$ $y = 80$	h. $y \times 7 = 35$ $y - 5$
i. $300 \times y = 1,200$ $y = 4$	j. $y \times 2 = 40$ $y = 20$

Multiplication Tables Review, p. 87

1.

$1 \times 5 = 5$	$7 \times 5 = 35$	$1 \times 10 = 10$	$7 \times 10 = 70$	$1 \times 11 = 11$	$7 \times 11 = 77$
$2 \times 5 = 10$	$8 \times 5 = 40$	$2 \times 10 = 20$	$8 \times 10 = 80$	$2 \times 11 = 22$	$8 \times 11 = 88$
$3 \times 5 = 15$	$9 \times 5 = 45$	$3 \times 10 = 30$	$9 \times 10 = 90$	$3 \times 11 = 33$	$9 \times 11 = 99$
$4 \times 5 = 20$	$10 \times 5 = 50$	$4 \times 10 = 40$	$10 \times 10 = 100$	$4 \times 11 = 44$	$10 \times 11 = 110$
$5 \times 5 = 25$	$11 \times 5 = 55$	$5 \times 10 = 50$	$11 \times 10 = 110$	$5 \times 11 = 55$	$11 \times 11 = 121$
$6 \times 5 = 30$	$12 \times 5 = 60$	$6 \times 10 = 60$	$12 \times 10 = 120$	$6 \times 11 = 66$	$12 \times 11 = 132$

10, 20, 30, 40, 50, 60. Because $10 = 2 \times 5$, so any product of 10 also is a product of 5.

2. For example, $50 = 5 \times 10 = 5 \times 2 \times 5$.

$1 \times 2 = 2$	$7 \times 2 = 14$	$1 \times 4 = 4$	$7 \times 4 = 28$	$1 \times 8 = 8$	$7 \times 8 = 56$
$2 \times 2 = 4$	$8 \times 2 = 16$	$2 \times 4 = 8$	$8 \times 4 = 32$	$2 \times 8 = 16$	$8 \times 8 = 64$
$3 \times 2 = 6$	$9 \times 2 = 18$	$3 \times 4 = 12$	$9 \times 4 = 36$	$3 \times 8 = 24$	$9 \times 8 = 72$
$4 \times 2 = 8$	$10 \times 2 - 20$	$4 \times 4 = 16$	$10 \times 4 = 40$	$4 \times 8 = 32$	$10 \times 8 = 80$
$5 \times 2 = 10$	$11 \times 2 = 22$	$5 \times 4 = 20$	$11 \times 4 = 44$	$5 \times 8 = 40$	$11 \times 8 = 88$
$6 \times 2 = 12$	$12 \times 2 = 24$	$6 \times 4 = 24$	$12 \times 4 = 48$	$6 \times 8 = 48$	$12 \times 8 = 96$

8, 16, 24. Because $8 = 2 \times 4$, so any product of 8 also is a product with 2 or 4.
For example, $32 = 4 \times 8 = 4 \times 2 \times 4$.

3.

$1 \times 3 = 3$	$7 \times 3 = 21$	$1 \times 6 = 6$	$7 \times 6 = 42$	$1 \times 9 = 09$	$7 \times 9 = 63$
$2 \times 3 = 6$	$8 \times 3 = 24$	$2 \times 6 = 12$	$8 \times 6 = 48$	$2 \times 9 = 18$	$8 \times 9 = 72$
$3 \times 3 = 9$	$9 \times 3 = 27$	$3 \times 6 = 18$	$9 \times 6 = 54$	$3 \times 9 = 27$	$9 \times 9 = 81$
$4 \times 3 = 12$	$10 \times 3 = 30$	$4 \times 6 = 24$	$10 \times 6 = 60$	$4 \times 9 = 36$	$10 \times 9 = 90$
$5 \times 3 = 15$	$11 \times 3 = 33$	$5 \times 6 = 30$	$11 \times 6 = 66$	$5 \times 9 = 45$	$11 \times 9 = 99$
$6 \times 3 = 18$	$12 \times 3 = 36$	$6 \times 6 = 36$	$12 \times 6 = 72$	$6 \times 9 = 54$	$12 \times 9 = 108$

In the table of nine, if you look at the number of tens each product has (colored red), they go from 0 to 9 in order, then repeat 9, then go to 10 (in 108), and would continue in order.
The ones digits start at 9 and decrease one by one to 0, then start over again.

6, 12, 18, 24, 30, 36. Because $6 = 2 \times 3$, so any product of 6 also is a product of 3. For example, $30 = 5 \times 6 = 5 \times 2 \times 3$.

4.

$1 \times 7 = 7$	$7 \times 7 = 49$	$1 \times 12 = 12$	$7 \times 12 = 84$
$2 \times 7 = 14$	$8 \times 7 = 56$	$2 \times 12 = 24$	$8 \times 12 = 96$
$3 \times 7 = 21$	$9 \times 7 = 63$	$3 \times 12 = 36$	$9 \times 12 = 108$
$4 \times 7 = 28$	$10 \times 7 = 70$	$4 \times 12 = 48$	$10 \times 12 = 120$
$5 \times 7 = 35$	$11 \times 7 = 77$	$5 \times 12 = 60$	$11 \times 12 = 132$
$6 \times 7 = 42$	$12 \times 7 = 84$	$6 \times 12 = 72$	$12 \times 12 = 144$

5. a. 7, 4, 8   b. 8, 5, 9   c. 8, 6, 7   d. 8, 6, 7   e. 9, 7, 8   f. 9, 6, 7
   g. 7, 9, 8   h. 9, 8, 6   i. 5, 9, 3   j. 12, 5, 6   k. 6, 12, 7   l. 9, 2, 4

6.

×	0	1	2	3	4	5	6	7	8	9	10	11	12
0	0	0	0	0	0	0	0	0	0	0	0	0	0
1	0	1	2	3	4	5	6	7	8	9	10	11	12
2	0	2	4	6	8	10	12	14	16	18	20	22	24
3	0	3	6	9	12	15	18	21	24	27	30	33	36
4	0	4	8	12	16	20	24	28	32	36	40	44	48
5	0	5	10	15	20	25	30	35	40	45	50	55	60
6	0	6	12	18	24	30	36	42	48	54	60	66	72
7	0	7	14	21	28	35	42	49	56	63	70	77	84
8	0	8	16	24	32	40	48	56	64	72	80	88	96
9	0	9	18	27	36	45	54	63	72	81	90	99	108
10	0	10	20	30	40	50	60	70	80	90	100	110	120
11	0	11	22	33	44	55	66	77	88	99	110	121	132
12	0	12	24	36	48	60	72	84	96	108	120	132	144

7. $90 - 8 \times 7 = 34$. So, 34 students went by bus.

8. a. 33   b. 26   c. 8   d. 70   e. 63   f. 26

9. a. 3   b. 9   c. 5   d. 4   e. 10

Scales Problems, p. 90

1. a. 7   b. 6   c. 6   d. 2   e. 7   f. 5.5   g. 2   h. 7

2. a. 11   b. 8   c. 13   d. 5

3. a. Hint: from the first balance we learn that one circle + one rectangle weigh 14. In the second balance, we have two circles and one rectangle, and of those, one rectangle and one circle together still weigh 14. This means that the one circle must weigh 6.
   Solution: 1 rectangle weighs 8 and 1 circle weighs 6.

   b. Hint: guess and check using small numbers. Look at the second scales, and in it, try for example one circle being 2. Then you will notice that that will not work! The circle has to be a bigger number. Check to see if the circle is 4. Then the diamond would be 0. That will not work either. Check to see if the circle is 6. Then the diamond would be 4. Now go to the first scale and check, if the circle = 6 and the diamond = 4 works there. If not, change your guess.
   Solution: 1 circle weighs 5 and 1 diamond weighs 2.

4. a. 1 circle weighs 4 and 1 square weighs 17.   b. 1 square weighs 5 and 1 triangle weighs 3.
   c. 1 square weighs 3 and 1 circle weighs 11.   d. 1 circle weighs 2 and 1 triangle weighs 3.

5. a. 70   b. 29   c. 60   d. 70   e. 9   f. 21   g. 18   h. 17   i. 8

Multiplying by Whole Tens and Hundreds, p. 94

1. a. 3,150; 35,600; 3,500   b. 620,000; 12,000; 13,000   c. 250,000; 38,000; 50,000

2. a. 160; 80; 100   b. 1400; 1,000; 2,200   c. 240; 700; 1,800   d. 320; 8,400; 1,080

3.

a. 20×7	b. 20×5	c. 200×8	d. 200×25
$= \underline{10} \times 2 \times 7$	$= 10 \times 2 \times 5$	$= 100 \times 2 \times 8$	$= 100 \times 2 \times 25$
$= 10 \times \underline{14}$	$= 10 \times 10$	$= 100 \times 16$	$= 100 \times 50$
$= \underline{140}$	$= 100$	$= 1600$	$= 5000$

4. $A = 20 \text{ ft} \times 15 \text{ ft} = 300 \text{ ft}^2$

5. $A = 15 \text{ ft} \times 200 \text{ ft} = 3,000 \text{ ft}^2$

6. One truckload costs $5 \times \$20 + \$30 = \$130$. Four truckloads cost $4 \times \$130 = \$520$.

7. a. 120; 160   b. 420; 550   c. 720; 450   d. 660; 480   e. 1,800; 2,800
   f. 4,200; 6,600   g. 2,400; 4,500   h. 3,300; 7,200   i. 1,320; 2400

8. a. 1,800; 21,000   b. 4,800; 27,000   c. 20,000; 40,000   d. 64,000; 100,000   e. 10,000; 1,200   f. 240,000; 99,000

9. One hour has _60_ minutes.
   How many minutes are in 12 hours? _12×60 min = 720 minutes_
   How many minutes are in 24 hours? _2×720 min = 1,440 minutes (double the previous result)_

10. One hour has _60_ minutes, and one minute has _60_ seconds.
    How many seconds are there in one hour? _60×60 sec = 3,600 seconds_

11. a. _$8 \times \$30 = \$240$_   b. _$40 \times \$30 = \$1,200$_

    c. Guess and check: _$2 \times \$240 = \$480$; $4 \times \$240 = \$960$; $5 \times \$240 = \$1,200$; so five days._

12. a. 120; 9   b. 8; 120   c. 10; 90   d. 160; 9   e. 50; 700   f. 70; 600

Puzzle Corner: $600 \times 50 = 6 \times 100 \times 5 \times 10 = 100 \times 10 \times (6 \times 5) = 1000 \times 30 = 30,000$.

Multiply in Parts 1, p. 98

1. a. 6×20 and 6×7. 120 and $42 = 162$ b. $(80 + 3)$; 5×80 and 5×3; 400 and $15 = 415$

   c. $(30 + 4)$; 9×30 and 9×4; 270 and $36 = 306$ d. $3 \times 99 = 3 \times 90$ and $3 \times 9 = 270 + 27 = 297$

   e. $7 \times 65 = 7 \times 60$ and $7 \times 5 = 420 + 35 = 455$ f. $4 \times 58 = 4 \times 50$ and $4 \times 8 = 200 + 32 = 232$

2. a. $9 \times 17 = 9 \times 10 + 9 \times 7 = 90 + 63 = 153$ b. $6 \times 29 = 6 \times 20 + 6 \times 9 = 120 + 54 = 174$
   c. $7 \times 33 = 7 \times 30 + 7 \times 3 = 210 + 21 = 231$

3.

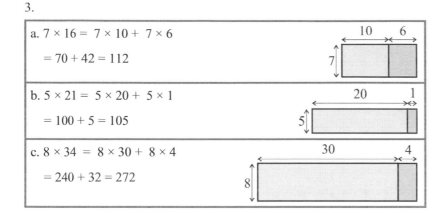

a. $7 \times 16 = 7 \times 10 + 7 \times 6$ $= 70 + 42 = 112$	
b. $5 \times 21 = 5 \times 20 + 5 \times 1$ $= 100 + 5 = 105$	
c. $8 \times 34 = 8 \times 30 + 8 \times 4$ $= 240 + 32 = 272$	

4. b. $3 \times 70 = 210$; $3 \times 3 = 9$; $210 + 9 = 219$ c. $4 \times 60 = 240$; $4 \times 7 = 28$; $240 + 28 = 268$
   d. $5 \times 90 = 450$; $5 \times 2 = 10$; $450 + 10 = 460$ e. $9 \times 30 = 270$; $9 \times 3 = 27$; $270 + 27 = 297$
   f. $7 \times 40 = 280$; $7 \times 7 = 49$; $280 + 49 = 329$

5. a. 65   b. 135   c. 165   d. 168   e. 88   f. 357

6. a. >   b. >   c. >

7. a. Jack's total cost was $8 \times \$14 = \$112$. You can multiply in parts: $8 \times 14 = 8 \times 10 + 8 \times 4 = 80 + 32 = 112$.
   b. There were $9 \times 14 + 56 = 182$ seats in all.
   c. The cost of the hammer is $3 \times \$17 = \$51$.

Multiply in Parts 2, p. 101

1. a. 3×100 and 3×20 and 3×7; 300 and 60 and $21 = 381$

   b. $(200 + 40 + 3)$ 5×200 and 5×40 and 5×3; $1,000$ and 200 and $15 = 1,215$

   c. $(300 + 10 + 4)$ 7×300 and 7×10 and 7×4; 2100 and 70 and $28 = 2,198$

   d. $(6,000 + 500 + 7)$ $4 \times 6,000$ and 4×500 and 4×7; $24,000$ and 2000 and $28 = 26,028$

   e. $5 \times 4,000 + 5 \times 800 + 5 \times 10 + 5 \times 3$; $20,000 + 4,000 + 50 + 15 = 24,065$

   f. $7 \times 2,000 + 7 \times 600 + 7 \times 90 + 7 \times 3$; $14,000 + 4,200 + 630 + 21 = 18,851$

2.

a. 8×127 $= 8 \times 100 + 8 \times 20 + 8 \times 7$ $= 800 + 160 + 56 = 1,016$	
b. 6×245 $= 6 \times 200 + 6 \times 40 + 6 \times 5$ $= 1,200 + 240 + 30 = 1,470$	
c. 9×196 $= 9 \times 100 + 9 \times 90 + 9 \times 6$ $= 900 + 810 + 54 = 1,764$	

Multiply in Parts 2, continued

3.

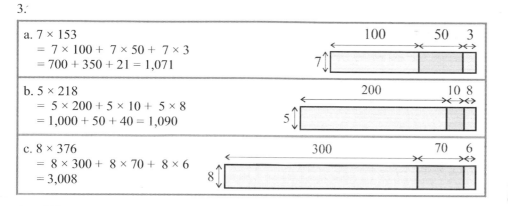

a. 7×153
   $= 7 \times 100 + 7 \times 50 + 7 \times 3$
   $= 700 + 350 + 21 = 1,071$

b. 5×218
   $= 5 \times 200 + 5 \times 10 + 5 \times 8$
   $= 1,000 + 50 + 40 = 1,090$

c. 8×376
   $= 8 \times 300 + 8 \times 70 + 8 \times 6$
   $= 3,008$

4. a. 512

b. $8 \times 100 = 800$; $8 \times 50 = 400$; $8 \times 1 = 8$; $800 + 400 + 8 = 1,208$

c. $3 \times 400 = 1,200$; $3 \times 50 = 150$; $3 \times 2 = 6$; $1,200 + 150 + 6 = 1,356$

d. $6 \times 3,000 = 18,000$; $6 \times 200 = 1,200$; $6 \times 10 = 60$; $6 \times 7 = 42$;

   $18,000 + 1,200 + 60 + 42 = 19,302$

e. $8 \times 2,000 = 16,000$; $8 \times 500 = 4,000$; $8 \times 50 = 400$; $8 \times 2 = 16$

   $16,000 + 4,000 + 400 + 16 = 20,416$

f. $6 \times 1,000 = 6,000$; $6 \times 90 = 540$; $6 \times 8 = 48$;

   $6,000 + 540 + 48 = 6,588$

5. a. $6 \times \$138 = \$600 + \$180 + \$48 = \$828$. In a year: $2 \times \$828 = \$1,656$
   b. 4×255 cm $= 800$ cm $+ 200$ cm $+ 20$ cm $= 1,020$ cm
   c. The bigger roll has 5×56 cm $= 250$ cm $+ 30$ cm $= 280$ cm of material.
   In total they have 56 cm $+ 280$ cm $- 336$ cm of material

6. b. 250¢ = \$2.50   c. 560¢ = \$5.60   d. 270¢ = \$2.70   e. 90¢ = \$0.90   f. 246¢ = \$2.46

7. a. \$12 + \$4.80 = \$16.80
   b. \$20 + \$3.50 = \$23.50
   c. $(4 \times \$12) + (4 \times \$0.50) = \$48 + \$2 = \$50$
   d. $(7 \times \$5) + (7 \times \$0.60) + (7 \times \$0.01) = \$35 + \$4.20 + \$0.07 = \$39.27$

8. a. $(5 \times \$2.00) + (5 \times \$0.70) = \$10.00 + \$3.50 = \$13.50$ total.
   b. Her change is $\$20.00 - \$13.50 = \$6.50$.

9. a. $(4 \times \$20.00) + (4 \times \$3.00) + (4 \times \$0.50) = \$80.00 + \$12.00 + \$2.00 = \$94.00$
   b. Their change is $\$100.00 - \$94.00 = \$6.00$.

More Practice, p. 105

1. a. $7 \times 900 + 7 \times 30 = 6,300 + 210 = 6,510$
   b. $7 \times 3,000 + 7 \times 400 + 7 \times 9 = 21,000 + 2,800 + 63 = 23,863$
   c. $6 \times \$11 + 6 \times \$0.80 + 6 \times 0.05 = \$66 + \$4.80 + \$0.30 = \$71.10$
   d. $5 \times \$2 + 5 \times \$0.90 + 5 \times \$0.03 = \$10 + \$4.50 + \$0.15 = \$14.65$
   e. $7 \times \$3 + 7 \times \$0.70 + 7 \times \$0.05 = \$21 + \$4.90 + \$0.35 = \$26.25$
   f. $8 \times \$10 + 8 \times \$0.90 + 8 \times \$0.05 = \$80 + \$7.20 + \$0.40 = \$87.60$

More Practice, continued

2. a. 4   b. 12   c. 20   d. 200   e. 9   f. 750

3. a. Susie orders 5 × 72 = 360 flowers in 5 weeks.
   b. It costs her 5 × $70 = $350 in five weeks

4.

a. Start at 80. Add 40 each time:	b. Start at 42,000. Subtract 3,000 each time:	c. Start at 1. Add 5 each time:
_120_	42,000	_1_
160	39,000	_6_
200	36,000	11
240	33,000	16
280	30,000	21
320	27,000	26
360	24,000	31
		36
		41
What does this pattern remind you of? The multiplication table of 4.	What does this pattern remind you of? The multiplication table of 3.	

5. When you add 5 to a number, it changes parity. In other words, if you add 5 to an odd number, you get an even number, and vice versa. In fact, the same happens when you repeatedly add any odd number.

Estimating in Multiplication, p. 107

1. Answers may vary. Estimating is not an "exact science".
   a. 5 × 70 = 350   b. 11 × 60 = 660   c. 120 × 8 = 960   d. 30 × 50 = 1,500   e. 7 × $4 = $28
   f. 8 × $12 = $96   g. 25 × $40 = $1,000   h. 9 × 20 − 180 or 10 × 17 =170   i. 60 × 900 = 54,000

2. a. 20 × $45 = $900; however since this is about shopping, it might be better not to round down so much, and estimate the cost as 24 × $45, which you can calculate in two parts: 20 × $45 and 4 × $45, which is $900 + $180 = $1,080.
   b. 500 × 20¢ = 10,000¢ = $100
   c. 200 × $1.50 = $200 + $100 = $300
   d. Tennis balls: 6 × $3 = $18; Rackets: 2 × $12 = $24; Total cost: $42.

3. a. Bill can buy 5 ads. Round to $350. Then add: two ads is $700, four ads is $1,400, six ads is $2,100. So he can not afford six ads. Five ads would be $1,400 + $350 = $1,750.
   b. Round the rate to $3 per hour. Since 8 × $3 = $24, she can rent them for about 8 hours.
   c. Beans: 8 × $0.30 = $2.40; Lentils: 5 × $0.40 = $2. So it is cheaper to buy 5 bags of lentils.
   d. Round the cost of string to $0.20 per foot. Round the number of children to 30. Cost per one child is about 8 × $0.20 = $1.60. The total cost is about 30 × $1.60 = $48.

Multiply in Columns—the Easy Way, p. 109

1. a. 456   b. 445   c. 301   d. 312   e. 415   f. 564   g. 288   h. 287

2. a. 822   b. 872   c. 2,191   d. 2,256   e. 3,608   f. 2,085
   g. 6,240   h. 1,944   i. 5,562   j. 1,698   k. 2,056   l. 3,040

3. a. 6   b. 90   c. 90

4. a. 581   b. 565

5. a. (236 − $40) × 7 = $196 × 7 = $1,372   b. 992 ft

6. a. 40   b. 40   c. 700   d. 12   e. 100   f. 80

Puzzle corner. a. 172; 1204   b. 358; 32   c. 709; 00

Multiply in Columns—the Easy Way, part 2, p. 112

1. a. 10,628   b. 30,528   c. 24,318   d. 56,712   e. 11,212   f. 26,460

2. a. 9,930   b. 18,495   c. 22,960   d. 36,702   e. 12,445   f. 34,408

3. a. The total distance is $2 \times 2 \times 3,820 = 4 \times 3,820 = 15,280$ feet.
   b. $55 + 4 \times 55 = 275$ marbles in total.

4. a. $5.49   b. $35.48   c. $80.50   d. $61.50
   e. $60.30   f. $58.94   g. $82.60   h. $318.48

5. a. Estimate: $10 \times 1.57 = 15.70$ or $10 \times 1.60 = 16.00$. Total cost: $9 \times \$1.57 = \$14.13$
   b. Estimate: $8 \times 2.3 = 18.40$. Change: $\$20 - 8 \times \$2.28 = \$20 - \$18.24 = \$1.76$

Multiply in Columns, the Standard Way, p. 115

1.

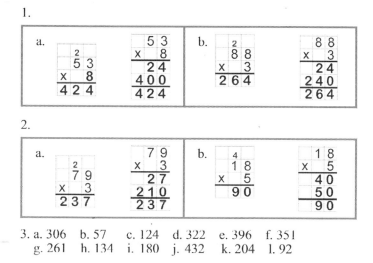

a.
```
   2
  5 3        5 3
 x   8      x   8
 4 2 4      2 4
            4 0 0
            4 2 4
```

b.
```
   2              8 8
  8 8           x   3
 x   3          2 4
 2 6 4          2 4 0
                2 6 4
```

2.

a.
```
   2
  7 9        7 9
 x   3      x   3
 2 3 7      2 7
            2 1 0
            2 3 7
```

b.
```
   4              1 8
  1 8           x   5
 x   5          4 0
 9 0            5 0
                9 0
```

3. a. 306   b. 57   c. 124   d. 322   e. 396   f. 351
   g. 261   h. 134   i. 180   j. 432   k. 204   l. 92

4. a. Three chairs: $3 \times \$48 = \$144$. Six chairs: $2 \times \$144 = \$288$ (double the previous result).
   b. Guess and check: $7 \times \$77 = \$539$ and $8 \times \$77 = \$616$. So, you need to work 8 days.

5.

a.
```
  1 2 3          1 2 3
 x     8        x     8
 9 8 4          2 4
                1 6 0
                8 0 0
                9 8 4
```

b.
```
  2 2            2 7 9
 2 7 9          x     3
 x     3        2 7
 8 3 7          2 1 0
                6 0 0
                8 3 7
```

c.
```
  4 6 3          4 6 3
 x     5        x     5
 2 3 1 5        1 5
                3 0 0
                2 0 0 0
                2 3 1 5
```

d.
```
  1 5 6          1 5 6
 x     6        x     6
 9 3 6          3 6
                3 0 0
                6 0 0
                9 3 6
```

6. a. 924   b. 3,542   c. 1,390   d. 2,233   e. 864   f. 7,281
   g. 861   h. 734   i. 10,872   j. 7,542   k. 24,708   l. 47,970

7. a. ___ $\times 43 = 304$ or more. Guess and check: $9 \times 43 = 387$; $8 \times 43 = 344$; $7 \times 43 = 301$. So, eight buses are needed.

   b. There are $304 + 24 = 328$ people. Eight buses have 344 seats. $344 - 328 = 16$. So, 16 seats are left empty.

Multiplying in Columns, Practice, p. 119

1. a. 387   b. Estimation: $8 \times 70 = 560$; Exact: 576
   c. Estimation: $6 \times 130 = 780$; Exact: 756
   d. Estimation: $5 \times 800 = 4,000$; Exact: 4,095
   e. Estimation: $3 \times 770 = 2,310$; Exact: 2,313
   f. Estimation: $6 \times 6,000 = 36,000$; Exact: 34,140
   g. Estimation: $4 \times 2,500 = 10,000$; Exact: 10,084
   h. Estimation: $3 \times 9,000 = 27,000$; Exact: 26,136

2. a. Estimation: $5 \times 200 = 1,000$; Exact: 980
   b. Estimation: $9 \times 200 = 1,800$; Exact: 1,845
   c. Estimation: $7 \times 400 = 2,800$; Exact: 2,716
   d. Estimation: $6 \times 700 = 4,200$; Exact: 4,266
   e. Estimation: $9 \times 10,000 = 90,000$; Exact: 88,263
   f. Estimation: $3 \times 3,000 = 9,000$; Exact: 8,943
   g. Estimation: $4 \times 6,000 = 24,000$; Exact: 22,152
   h. Estimation: $6 \times 5,000 = 30,000$; Exact: 28,860

3. a. It has $3 \times 187 = 561$ pages.
   b. It takes one hour to drive 60 km, and $10 \times 60 = 600$ km, so it takes 10 hours to drive 600 km.

4. a. 80   b. 3   c. 30

5. a. $5 \times 250 = 1,250$, which is not enough. $6 \times 250 = 1,500$, so she needs to buy six packages.
   b. Perimeter: 2×9 ft $+ 2 \times 28$ ft $= 18$ ft $+ 56$ ft $= 74$ ft.
     Area: 9 ft $\times$ 28 ft $= 252$ sq. ft.

6. a. Every time she multiplies she carries the ones digit instead of the tens digit.
     The real answers are: 234, 342, 532.
   b. Andy does not do the carries. He writes the tens digit he should carry as part of the answer.
     The real answers are: 84, 225, 645.

Puzzle Corner: 117; $174 \times 5 = 870$; $138 \times 7 = 966$; $3,219 \times 3 = 9657$

Order of Operations Again, p. 123

1. a. First calculate the sum $\underline{23 + 31}$.
   b. Next multiply that sum by $\underline{9}$.
   c. Subtract that result from $\underline{650}$.
   d. Lastly add $\underline{211}$ to the subtraction result.
   Result: 375

2. a. 1,800; 150   b. 210; 2,000   c. 3,200; 0   d. 2,800; 5,600

3. a. 100; 900   b. 4,100; 100   c. 1,000; 400   d. 900; 860

4. a. 600; 1,400   b. 3,500; 700   c. 230; 2,660   d. 2,700; 5,800

5. a. $N = 30$ b. $N = 3$ c. $N = 3$ d. $N = 12$

6. a. 316   b. 2,155   c. 1,124

7. a. $4 \times \$2 + 3 \times \$3 = \$17$ b. $4 \times (\$2 + \$3) = \$20$ c. $\$50 - 5 \times \$3 - 5 \times \$2 = \$25$

8. a. $7 \times \$2.55 = \$17.85$. $\$20 - \$17.85 = \$2.15$
   b. 8×3 kg $+ 15 \times 2$ kg $= 54$ kg
   c. The first building is 9×9 ft $= 81$ ft tall. The second one is 3×81 ft $= 243$ feet tall.
   d. $6 \times \$4.25 + 3 \times \$8.50 = \$25.50 + \$25.50 = \$51$
   e. $5 \times \$3.60 = \$18$ and $4 \times \$4.80 = \$19.20$. Four meters of material for \$4.80 per meter costs more.

Puzzle Corner: a. $7 \times (2 + 8) = 70$ b. $80 - 5 \times (10 - 5) = 55$ c. $(4 + 8) \times 5 - 20 = 40$

1. a. Estimation: $4 \times \$5 = \$20$; answer $18.20   b. Estimation: $4 \times \$10 = \$40$; answer $38.80
   c. Estimation: $7 \times \$5 = \$35$; answer $34.37   d. Estimation: $6 \times \$1 = \$6$; answer $4.92
   e. Estimation: $7 \times \$13 = \$91$; answer $87.71   f. Estimation: $5 \times \$43 = \$215$; answer $215.75

2. Estimates vary. For example: $8 \times \$1.40 = \$11.20$. Change: $8.80.
   Solution: $8.88. Calculations: Multiply in columns or in parts $8 \times \$1.39 = \$11.12$. Then, subtract $\$20 - \$11.12 = \$8.88$

3.

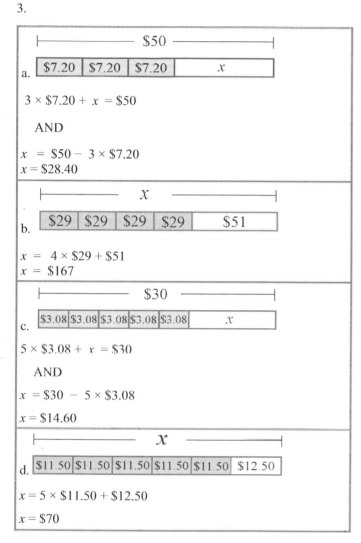

a.
$3 \times \$7.20 + x = \$50$

AND

$x = \$50 - 3 \times \$7.20$
$x = \$28.40$

b.
$x = 4 \times \$29 + \$51$
$x = \$167$

c.
$5 \times \$3.08 + x = \$30$

AND

$x = \$30 - 5 \times \$3.08$
$x = \$14.60$

d.
$x = 5 \times \$11.50 + \$12.50$
$x = \$70$

4. Notice that you need <u>five</u> packs because five packs will contain 20 bottles.
   Estimation: $5 \times \$3 = \$15$.
   Calculation: $5 \times \$2.76 = \$13.80$.

5. Estimation: $20 \times \$0.15 = 300$ cents $= \$3$; $10 \times \$1 = \$10$; Total $13.
   Calculation: $20 \times \$0.15 = 300$ cents $= \$3$. The student can solve $10 \times \$1.09$ by multiplying in parts: $10 \times \$1$ is $10,
   and 10×9 cents $= 90$ cents. The notebooks thus cost $10.90, and the total cost is $13.90.

6. Estimation: $4 \times \$10 + \$65 + \$26 = \$40 + \$65 + \$26 = \$131$.
   Calculations: $4 \times \$9.80 = \$39.20$. $\$39.20 + \$65 + \$25.80 = \$130$

Puzzle corner. $20. The price of one wheelbarrow was $125 because $8 \times \$125 = \$1000$.

1. a.

Miles	45	90	135	180	225	270	315	360	405	450
Hours	1	2	3	4	5	6	7	8	9	10

b.

Dollars	$5.10	$10.20	$15.30	$20.40	$25.50	$30.60	$35.70	$40.80	$45.90	$51.00
Meters	1	2	3	4	5	6	7	8	9	10

c.

Dollars	$1.50	$3.00	$4.50	$6.00	$7.50	$9.00	$10.50	$12.00	$13.50	$15.00
Cans	1	2	3	4	5	6	7	8	9	10

d.

Dollars	$30	$60	$90	$120	$150	$180	$210	$240	$270	$300
Buckets	2	4	6	8	10	12	14	16	18	20

e.

Feet	40	80	120	160	200	240	280	320	360	400
Minutes	10	20	30	40	50	60	70	80	90	100

f.

Tires	Minutes
1	15
2	30
3	45
4	60
5	75
6	90

g.

Days	Scarves
3	1
6	2
9	3
12	4
15	5
18	6

h.

Hours	Dollars
1	$15
2	$30
3	$45
4	$60
5	$75
6	$90

Two sacks of the same size contain 12 kg of potatoes total. How many of that size of sack would you need to get 30 kg of potatoes?

To solve the problem, make up a little table as on the right:

1 sack	_6_ kg
2 sacks	12 kg
3 sacks	3 × 6 kg = 18 kg
_5_ sacks	_5_ × 6 kg = 30 kg

The total number of chocolates in five identical boxes was 30. How many chocolates would two boxes contain?

First find out how many in ONE box:

1 box	_6_ chocolates
2 boxes	_12_ chocolates
5 boxes	30 chocolates

So Many of the Same Thing, continued

2. a.

1 flower	$3
5 flowers	$15
6 flowers	$18

b.

1 can	200 g
3 cans	600 g
4 cans	800 g

c.

1 lure	$2
3 lures	$6
7 lures	$14

d.

1 episode	30 min
3 episodes	90 min
5 episodes	150 min

e.

1 sit-up	2 sec
5 sit-ups	10 sec
30 sit-ups	60 sec

f.

1 notebook	$2
7 notebooks	$14
10 notebooks	$20

g.

1 day	30 pages
4 days	120 pages
10 days	300 pages

h.

1 pair	$0.75
6 pairs	$4.50
30 pairs	$22.50

3. a.

5 cars	$35.50
1 car	$7.10
4 cars	$28.40

b.

5 rows	100 minutes
1 row	20 minutes
9 rows	180 minutes = 3 hours

c. 25 minutes.
Solution: Elaine can run 4 times around a track in an hour. This means she takes 15 minutes to run around the track. Today she ran three times. She took 45 minutes for that. Then walked the fourth time. All that took 10 minutes longer than on her normal days. This means she took 1 hour 10 minutes. So, if running took 45 minutes, and in total she used 1 h 10 min, then the walking time is the difference of those, which is 25 minutes.

Multiplying Two-Digit Numbers in Parts, p. 131

1. a. $23 \times 31 = 20 \times 30 + 20 \times 1 + 3 \times 30 + 3 \times 1 = 600 + 20 + 90 + 3 = 713$ square units.
   b. $28 \times 45 = 20 \times 40 + 20 \times 5 + 8 \times 40 + 8 \times 5 = 800 + 100 + 320 + 40 = 1,260$ square units.
   c. $35 \times 27 = 30 \times 20 + 30 \times 7 + 5 \times 20 + 5 \times 7 = 600 + 210 + 100 + 35 = 945$ square units.

2.

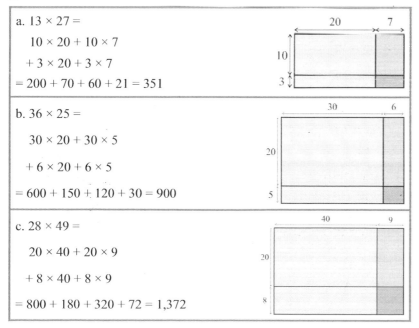

a. $13 \times 27 =$
  $10 \times 20 + 10 \times 7$
  $+ 3 \times 20 + 3 \times 7$
  $= 200 + 70 + 60 + 21 = 351$

b. $36 \times 25 =$
  $30 \times 20 + 30 \times 5$
  $+ 6 \times 20 + 6 \times 5$
  $= 600 + 150 + 120 + 30 = 900$

c. $28 \times 49 =$
  $20 \times 40 + 20 \times 9$
  $+ 8 \times 40 + 8 \times 9$
  $= 800 + 180 + 320 + 72 = 1,372$

3.

	a.			b.	
		8 7			2 4
		× 1 5			× 7 1
5 × 7 →		3 5	1 × 4 →		4
5 × 80 →		4 0 0	1 × 20 →		2 0
10 × 7 →		7 0	70 × 4 →		2 8 0
10 × 80 →		+ 8 0 0	70 × 20 →		+ 1 4 0 0
		1 3 0 5			1 7 0 4

	c.			d.	
		3 8			5 2
		× 9 2			× 6 5
2 × 8 →		1 6	5 × 2 →		1 0
2 × 30 →		6 0	5 × 50 →		2 5 0
90 × 8 →		7 2 0	60 × 2 →		1 2 0
90 × 30 →		+ 2 7 0 0	60 × 50 →		+ 3 0 0 0
		3 4 9 6			3 3 8 0

4.

a.	5 5	b.	8 1	c.	7 3	d.	9 9
	× 1 2		× 6 4		× 8 0		× 1 1
	1 0		4		0		9
	1 0 0		3 2 0		0		9 0
	5 0		6 0		2 4 0		9 0
	+ 5 0 0		+ 4 8 0 0		+ 5 6 0 0		+ 9 0 0
	6 6 0		5 1 8 4		5 8 4 0		1 0 8 9

5.

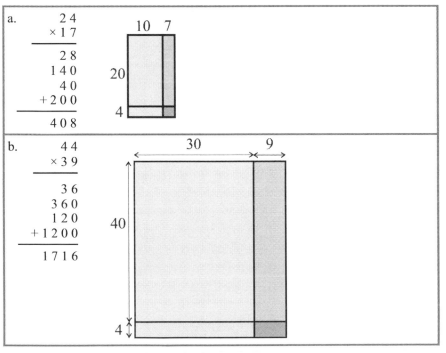

a.
```
   2 4
 × 1 7
 ─────
   2 8
 1 4 0
   4 0
+2 0 0
 ─────
 4 0 8
```

b.
```
   4 4
 × 3 9
 ─────
   3 6
 3 6 0
 1 2 0
+1 2 0 0
 ─────
 1 7 1 6
```

6.

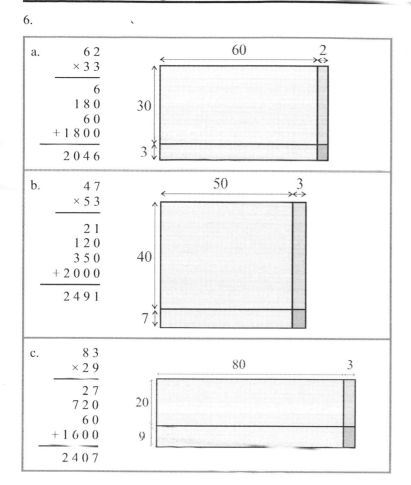

a.
```
     6 2
   × 3 3
   ─────
       6
     1 8 0
       6 0
   + 1 8 0 0
   ─────────
   2 0 4 6
```

b.
```
     4 7
   × 5 3
   ─────
       2 1
     1 2 0
     3 5 0
   + 2 0 0 0
   ─────────
   2 4 9 1
```

c.
```
     8 3
   × 2 9
   ─────
       2 7
     7 2 0
       6 0
   + 1 6 0 0
   ─────────
   2 4 0 7
```

Multiply by Whole Tens in Columns, p. 136

1. a. 5,220   b. 2,040 (first multiply 51 × 4)   c. 1,980 (First multiply 66 × 3)

2. a. 20 × 65 kg = 1,300 kg   b. 4 × 25 = 100 apples in each crate   c. He got three crates, which is 300 apples

3. a. 3,680   b. 1,620   c. 24,150   d. 7,160   e. 10,980

4. a. 3,200   b. 8,400   c. 18,480   d. 46,060

5. 5 × 250 km = 1,250 km;   4 × 1,250 km = 5,000 km

6. 7 × 800 m = 5,600 m or 5 km 600 m

7. a. 19,480   b. 43,500   c. 6,553

Puzzle Corner: 1,107,800 (1 million 107 thousand 800)

Multiplying in Parts: Another Way, p. 138

1. b. 40×73 and 8×73 c. 10×42 and 9×42 d. 50×89 and 5×89

2. a. $20 \times 16 = 320$; $8 \times 16 = 128$; $320 + 128 = 448$ b. $40 \times 73 = 2,920$; $8 \times 73 = 584$; $2,920 + 584 = 3,504$
   c. $10 \times 42 = 420$; $9 \times 42 = 378$; $420 + 378 = 798$ d. $50 \times 89 = 4,450$; $5 \times 89 = 445$; $4,450 + 445 = 4,895$

3. a. 40×41 and 6×41
      $40 \times 41 = 1,640$; $6 \times 41 = 246$; $1,640 + 246 = 1,886$

   b. 20×39 and 8×39
      $20 \times 39 = 780$; $8 \times 39 = 312$; $780 + 312 = 1,092$

   c. $10 \times 27 + 5 \times 27$
      $10 \times 27 = 270$; $5 \times 27 = 135$; $270 + 135 = 405$

   d. $90 \times 16 + 3 \times 16$
      $90 \times 16 = 1,440$; $3 \times 16 = 48$; $1,440 + 48 = 1,488$

4. $12 \times \$27 = \$324$. To multiply in parts, multiply $10 \times \$27 = \$270$, then $2 \times \$27 = \$54$, and add those.
   Or, solve it this way: for two months, the bill is $54. For four months, it is $108. Then multiply that times 3 to get $324.
   Estimates vary. For example, $12 \times \$30 = \$360$.

5. 720 Estimates vary. For example, write $1 \times 2 \times 3 \times 4 \times 5 \times 6$ as 24×30.
   Estimate that as $20 \times 30 = 600$, or as $25 \times 30 = 750$.

6. 12×15 kg $= 180$ kg. To multiply in parts, multiply either 10×15 and 2×15, OR 10×12 and 5×12.
   Estimates vary. For example, 10×15 kg $= 150$ kg.

7. $12 \times \$35 = \$420$. To multiply in parts, multiply either 10×35 and 2×35, OR 30×12 and 5×12.
   Estimates vary. For example, $10 \times \$35 = \$350$, or $12 \times \$40 = \$480$.

The Standard Multiplication Algorithm with a Two-Digit Multiplier, p. 140

1. a. $8 \times 65 = 520$. $520 + 650 = 1,170$ b. $4 \times 82 = 328$. $328 + 7,380 = 7,708$
   c. $20 \times 93 = 1,860$. $1,860 + 186 = 2,046$ d. $50 \times 70 = 3,500$. $3,500 + 210 = 3,710$

2. a. 636   b. 385   c. 6,396   d. 980   e. 494   f. 950   g. 884   h. 931

3. a. Estimate: $60 \times 10 = 600$. Answer 728
   b. Estimate: $30 \times 60 = 1,800$ OR $20 \times 70 = 1,400$. Answer: 1,625. Here, it helps to round one factor down and one
      up, since both numbers end in 5. (If you round both up, you get $30 \times 70 = 2,100$, which is off a lot from the answer).
   c. Estimate: $70 \times 40 = 2,800$. Answer: 2,860
   d. Estimate: $45 \times 10 = 450$. Answer: 616
   e. Estimate: $90 \times 50 = 4,500$. Answer: 4,232
   f. Estimate: $90 \times 80 = 7,200$. Answer: 7,161

4. a. Estimate: $80 \times 80 = 6,400$. Answer: 6,561   b. Estimate: $100 \times 30 = 3,000$. Answer: 3,201
   c. Estimate: $30 \times 50 = 1,500$. Answer: 1,512

5. a. $15 \times 12 = 180$ eggs. Estimate: $15 \times 10 = 150$
   b. $21 \times 60 = 1,260$ minutes. Estimate: $20 \times 60 = 1,200$
   c. $11 \times 39 = 429$. No, 11 buses are not enough. Estimate: $11 \times 40 = 440$.
   d. $12 \times \$21 = \$252$. Estimate: $12 \times \$20 = \$240$.

6. a. Estimate: $50 \times 20 = 1,000$. Answer: 969   b. Estimate: $45 \times 30 = 1,350$. Answer: 1,260
   c. Estimate: $10 \times 20 = 200$. Answer: 216   d. Estimate: $80 \times 100 = 8,000$. Answer: 7,980

7. a. Estimation: $\$300 - 15 \times \$20 = \$300 - \$300 = \$0.00$. Answer: $45
   b. Estimation: $52 \times \$100 = \$5,200$. Answer: $5,096

8. a. 3,600; 0   b. 300; 12,100

Puzzle Corner: $3 \times \underline{10}5 = 315$; $\underline{6} \times 6\underline{6}7 = 4002$; $4 \times \underline{234} = 9\underline{3}6$; $5 \times 8\underline{34} = 4,\underline{170}$

Mixed Review, p. 144

1. a. 1,400   b. 4,000   c. 5,200   d. 340   e. 200   f. 9,000

2.

a. $x = 126 - 57 = 69$	b. $x = 2,000 - 1,199 = 801$

3. a. 440,000   b. 220,000   c. 617,100   d. 200,000   e. 300,000   f. 81,000

4. a. 7,600; 4,000   b. 6,190; 26,700   c. 98,000; 430,000

5. a. 156; 80; 84   b. 70; 150; 464   c. 21; 9,990; 1,000

6. a. 119; 980   b. 0; 700

7. a. <   b. =   c. <

8. Each time, Mason forgets to add the regrouped tens digit. Correct answers: 336, 384, 717.

9. a. 3 × $345 + $345 = $1,380 total
   b. $145,600 + $12,390 = $157,990. $157,990 × 3 = $473,970 total costs for June, July, and August.
      No, the total cost is not more than half a million dollars.

Review, p. 146

1. a. 1,200; 180   b. 4,200; 3,300   c. 81,000; 40,000

2. a. 80; 7   b. 4; 400   c. 300; 80

3. a. 40   b. 2   c. 40

4. In about 8 weeks. 8 × $500 = $4,000.

5. a. Estimation: 7 × 50 = 350; Exact: 336
   b. Estimation: 6 × 800 = 4,800; Exact: 4,878
   c. Estimation: 20 × 20 = 400; Exact: 378
   d. Estimation: 4 × 6,000 = 24,000; Exact: 23,612

6.

Roses	1	2	3	4	5	6	7	8
Price	$0.90	$1.80	$2.70	$3.60	$4.50	$5.40	$6.30	$7.20

7. 2 × 98 = 196;   8 × 17 = 136;   196 − 136 = 60

8. a. 2,000   b. 0   c. 80   d. 20,000

9.

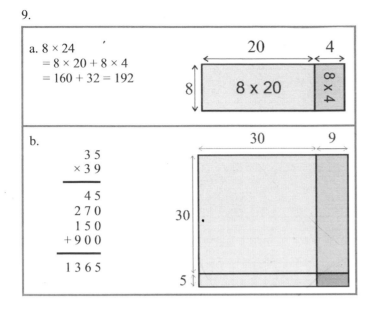

a. 8×24
    $= 8 \times 20 + 8 \times 4$
    $= 160 + 32 = 192$

b.

$$\begin{array}{r} 35 \\ \times 39 \\ \hline 45 \\ 270 \\ 150 \\ +900 \\ \hline 1365 \end{array}$$

10. a. He has $50 \times 20 = 1,000$ shirts. The cost is $1,000 \times \$2 = \$2,000$.

   Or, write a single number sentence $50 \times 20 \times \$2 = \$2,000$.

   b. $8 \times \$2.35 = \$18.80$; $\$20 - \$18.80 = \$1.20$. Or, $\$20 - 8 \times \$2.35 = \$1.20$

   c. $5 \times \$1.50 + \$12.50 = \$20$

   d. $6 \times \$1.28 + 7 \times \$2.15 = \$22.73$

   e. $\$45 - \$8 = \$37$. $5 \times \$37 = \$185$. Or, $5 \times (\$45 - \$8) = \$185$.

   f. 6 miles. Think first how many miles it runs in 5 minutes.
      You can make a table (see below):

Miles	Minutes
9	15
3	5
6	10

Chapter 4: Time and Measuring

Time Units, p. 152

1. a.

Days	Hours
1	24
2	48
3	72
4	96
5	120
6	144
7	168
8	192

b.

Minutes	Seconds
1	60
2	120
3	180
4	240
5	300
6	360
7	420
8	480

c.

Years	Months
1	12
2	24
3	36
4	48
5	60
6	72
7	84
8	96

2. a. $321   b. $1,095

3. a. 300 min; 600 min; 720 min
   b. 246 min; 217 min; 470 min
   c. 498 min; 1,210 min; 723 min

4. a. 8 hours 45 min   b. 10 hours 55 min   c. 7 hours 32 min   d. 15 hours 37 min

5. a. $7 \times 35 = 245$ min or 4 hours 5 min.
   b. 2 hours 30 min + 3 hours 50 min + 1 hour 10 min + 3 hours 25 min. = 10 hours 55 min.

6. a. 45 min + 35 min + 1 h 10 min + 1 h 5 min + 40 min = 4 hours 15 min.
   b. 2×40 min = 80 min or 1 hour 20 min. 1 hour 20 min + 3 hours = 4 hours 20 min total.
   c. 2 minutes = 120 seconds. Jean's finishing time was $120 - 24 = 96$ seconds or 1 min 36 seconds.
   d. 1 day = 24 hours. $3 \times 24 = 72$ hours.
   e. $7 \times 25 = 175$ minutes or 2 hours 55 min. In four weeks' time, you would walk your dog about $4 \times 175 = 700$ min or 11 hours 40 min.
   f. There are 60 seconds in one minute and 60 minutes in one hour. Multiply $60 \times 60 = 3,600$ seconds.

The 24-Hour Clock, p. 155

1. a. 5:40   b. 20:00   c. 18:15   d. 11:04   e. 12:30   f. 16:35   g. 23:55   h. 19:05

2. a. 3:00 p.m.   b. 5:29 p.m.   c. 4:23 a.m.   d. 11:55 p.m.   e. 2:30 p.m.   f. 10:45 a.m.   g. 4:00 p.m.   h. 9:15 p.m.

3. a. Bus 2   b. Bus 6   c. 1 h 20 min   d. 18:01, or 6:01 PM.   e. 33 minutes   f. 52 minutes
   g. Mark has to arrive at Newmarket at 6:15 at the latest, so Bus 8 that leaves at 17:25 will work.

Elapsed Time or How Much Time Passes, p. 157

1. a. 1 hr 30 min   b. 1 hr 40 min   c. 6 hr 41 min   d. 4 hr 40 min   e. 4 hr 2 min   f. 7 hr 30 min

2. a. 3 hr 33 min   b. 2 hr 26 min   c. 5 hr 29 min

3. a. 5 hr 45 min   b. 8 hr 16 min   c. 9 hr 38 min

4. a. 5 hr 18 min   b. 10 hr 40 min   c. 13 hr 40 min

Elapsed Time or How Much Time Passes, continued

5.

	Time
Patient 1	8:00 - 8:30
Patient 2	8:30 - 9:00
Patient 3	9:00 - 9:30
break	9:30 - 9:50
Patient 4	9:50 - 10:20
Patient 5	10:20 - 10:50
Patient 6	10:50 - 11:20
break	11:20 - 11:40

	Time
Patient 7	11:40 - 12:10
Patient 8	12:10 - 12:40
Patient 9	12:40 - 13:10
break	13:10 - 13:30
Patient 10	13:30 - 14:00
Patient 11	14:00 - 14:30
Patient 12	14:30 - 15:00

6.

Class	Time
Social Studies	8:00 - 8:50
Math	8:55 - 9:45
Science	9:50 - 10:40
English	10:45 - 11:35

Class	Time
Lunch	11:35 - 12:15
History	12:15 - 1:05
P.E.	1:10 - 2:00

7. Answers will vary. Please check the student's work.

8. a. 6:10   b. 15:25   c. 18:30   d. 11:15   e. 22:03   f. 6:22
   g. Shift 1: 8 hours 30 min; Shift 2: 8 hours; Shift 3: 9 hours. Each shift overlaps the next by 30 minutes.

9. a. 1:20 p.m.   b. 7:42 p.m.   c. 2:45 p.m.   d. 3:24 p.m.
   e. 10:40 a.m.   f. 1:22 a.m.   g. 14:45 p.m.   h. 13:50 p.m.

10. a. They left home at 7:15.
    b. They should leave the city by 16:45.

c.

	Mo	Wd	Th	Fr	Sa
Start: End:	17:15 18:20	17:03 18:05	17:05 18:12	17:45 18:39	17:12 18:15
Running time:	1 h 5 min	1 h 2 min	1 h 7 min	54 min	1 h 3 min

d. His total running time was 5 hours and 11 minutes.
   e. From 8:30 until 12:00 is 3 hours 30 min. Then, from 12:00 until
      17:15 is 5 hours 15 min. 3 hours 30 min + 5 hours 15 min is
      8 hours 45 min. Now, subtract the total amount of time, he
      had off for breaks: 8 hours 45 min − 60 minutes = 7 hours 45 min
      of actual work time.
   f. From 7:30 a.m. until 9 p.m. is 13 hours 30 min. 7 × 13 hours 30 min = 91 hours
      210 min = 94 hours 30 min a week.
   g. The flight was delayed 15 min, so it will arrive 15 min later than 5:10 p.m., at 5:25.

1.

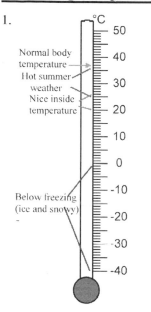

Normal body temperature →
Hot summer weather
Nice inside temperature
Below freezing (ice and snowy)

°C
50
40
30
20
10
0
-10
-20
-30
-40

2. a. 20 °C   b. 37°C   c. 12°C   d. 6°C   e. 29°C

3. Answers will vary.

4. Answers will vary.

5.

a fall day   5°C
a summer day   39°C
a fever   22°C
hot soup   55°C
boiling oil   -12°C
It is snowing!   200°C
inside a fridge   12°C
inside a house   21°C

6. a.

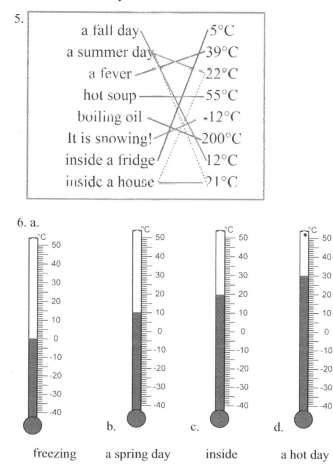

| freezing | b. a spring day | c. inside | d. a hot day |

7. a. -6°C   b. -3°C   c. -11°C   d. -16°C   e. -13°C

Measuring Temperature: Celsius, continued

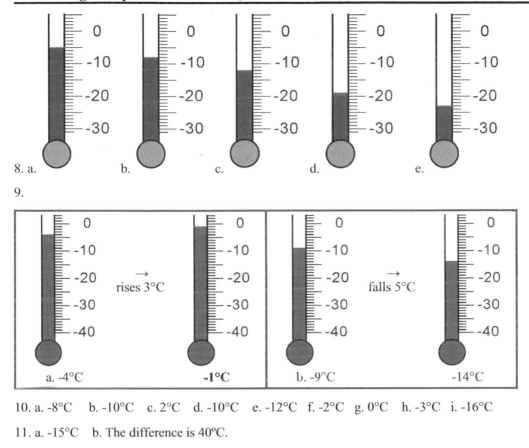

8. a.   b.   c.   d.   e.

9.

a. -4°C   rises 3°C → **-1°C**   b. -9°C   falls 5°C → -14°C

10. a. -8°C b. -10°C c. 2°C d. -10°C e. -12°C f. -2°C g. 0°C h. -3°C i. -16°C

11. a. -15°C   b. The difference is 40°C.

Puzzle corner: 2°C

Measuring Temperature: Fahrenheit, p. 166

1. a. 50°F; Chilly day   b. 81°F; Nice weather   c. 93°F; Very warm day   d. 105°F; Hot desert   e. 73°F; Inside

2. Answers will vary.

3. Answers will vary.

4. Answers will vary. For example:
   a. a very cool autumn morning;   b. a winter day;   c. a very hot summer day.

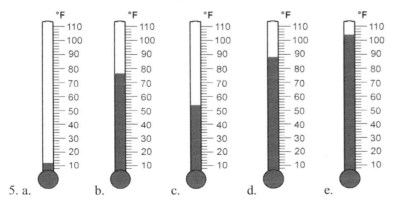

5. a.   b.   c.   d.   e.

Temperature Line Graphs, p. 168

Month	Jan	Feb	Mar	Apr	May	Jun	Jul	Aug	Sep	Oct	Nov	Dec
Max Temperature	6°C	7°C	10°C	13°C	17°C	20°C	22°C	21°C	19°C	14°C	10°C	7°C

1. a. July   b. January   c. March and November; or February and December.
   d. 3 degrees Celsius.   e. 2 degrees Celsius.   f. 16 degrees Celsius

2.

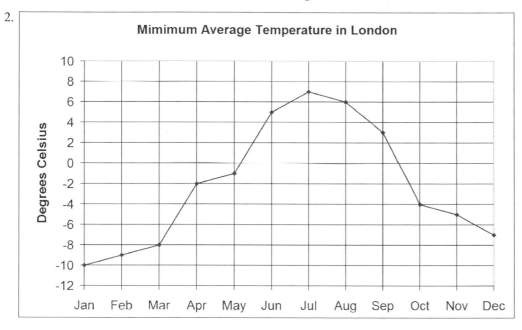

a. January   b. 6°   c. 1°   d. 15°

Measuring Length, p. 170

1. a. 1 3/4 in. or 4 cm 5 mm   b. 2 1/2 in or 6 cm 4 mm   c. 3 1/4 in. or 8 cm 3 mm
   d. 5 1/4 in. or 13 cm 4 mm   e. 4 1/2 in. or 11 cm 5 mm

2. a. 1 1/8 in   b. 2 3/8 in   c. 1 7/8 in   d. 7/8 in   e. 4 3/8 in   f. 3 3/4 in

3. Check students' work.

   a. ───────────────────

   b. ─────────────────────

   c. ──────────────────────────────────

   d. exceeds the width of this paper

   e. exceeds the width of this paper

4. Check students' work.

   a. ──────────

   b. ───────────────────────

   c. ─────────

   d. and e. exceed the width of this paper.

5. Answers will vary.

More Measuring in Inches and Centimeters, p. 173

1. Answers will vary.

2. Answers will vary.

3. a. 2 1/8 in, 2 1/4 in, 5 in, 3 1/2 in   b. 5 cm 3 mm, 5 cm 7 mm, 12 cm 8mm, 9 cm

4. a. 3 cm   b. 7 cm   c. 6 inches

5. 1 cm 4 mm

6. a. 10   b. 4   c. 4 1/2 inches   d. 1 1/2 inches   e. 3 inches

7. Answers will vary. Please check the student's work.

Feet, Yards, and Miles, p. 175

1. Answers will vary.

2. Check student's work.

3. Answers will vary.

4.

a.		b.		c.	
Feet	Inches	Feet	Inches	Feet	Inches
1	12	6	72	11	132
2	24	7	84	12	144
3	36	8	96	13	156
4	48	9	108	14	168
5	60	10	120	15	180

5. a. 72 in; 132 in   b. 29 in; 92 in   c. 163 in; 143 in   d. 3 ft; 4 ft 2 in   e. 2 ft 3 in; 8 ft 4 in   f. 5 ft 4 in; 7 ft 1 in

6. a. He is 8 inches taller.
   b. She is 5 ft 1 in. tall now.
   c. The difference between their heights is 6 ft 6 in.
   d. The shorter sides measure 2 ft 11 in.

7.

a.		b.		c.	
Yards	Feet	Yards	Feet	Yards	Feet
1	3	4	12	7	21
2	6	5	15	8	24
3	9	6	18	9	27

8. a. 18 ft; 39   b. 8 ft; 16 ft   c. 8 yd; 14 yd   d. 4 yd 1 ft; 5 yd 2 ft   e. 7 yd 1 ft; 9 yd 2 ft   f. 10 yd 2 ft; 13 yd 1 ft

9. 400 yard = 1,200 feet, so Jessie ran the longer distance. He ran 1,200 ft − 1,000 ft = 200 feet longer distance.

Feet, Yards, and Miles, continued

10. a. ? = 120 ÷ 2 − 18 = 42 yd.   b. 42 × 3 =126 ft

11. 6 yards = 18 ft, so 18 ft − 2 ft − 2 ft = 14 ft of material left, or 4 yd 2 ft.

12. The remaining piece is 1 ft 8 in long.

13 a. 18 ft 4 in   b. 2 ft 8 in   c. 8 ft 4 in
    d. 5 ft 6 in + 8 ft 3 in + 3 ft 8 in = 17 ft 5 in, so yes, they will fit.   e. It is 8 ft tall.

14. a. 21,120 ft   b. 28,750 ft   c. 13,000 ft is longer.   d. About 4 miles.
    e. About 5 1/2 miles.   f. 15,000 feet; about 3 miles.   g. He walks 10 × 950 ft = 9,500 ft; about two miles.

Metric Units for Measuring Length, p. 180

1. and 2. Answers will vary.

3. a. 500 cm; 800 cm; 1,200 cm   b. 406 cm; 919 cm; 1,080 cm   c. 8 m; 2 m 39 cm; 4 m 7 cm

4. a. 50 mm; 80 mm; 140 mm   b. 28 mm; 75 mm; 104 mm   c. 5 cm 0 mm; 7 cm 2 mm; 14 cm 5 mm

5. a. 5,000 m; 23,000 m; 1,200 m   b. 2,800 m; 6,050 m; 13,579 m   c. 2 km; 4 km 300 m; 18 km 700 m

6. a. 14 km 100 m   b. 6 km 400 m   c. 4 km 100 m   d. 4 km 200 m

7. a. 5 m 60 cm   b. 4 cm 4 mm   c. 22 cm 4 mm   d. 1 cm 5mm

8. a. 1,000 mm   b. 3 km 600 m in a day; 18 km in a week.   c. Jared is 8 cm taller.
    d. She can have 12 complete butterflies on a wall that is 1 meter long,
       and 37 complete butterflies on a wall that is 3 meters long.

Customary Units of Weight, p. 183

1. Answers will vary.

2. a. 1 oz   b. 1 lb or 20 oz   c. 4 T   d. 2 T or 3500 lb
   c. 5 oz   f. 130 lb   g. 3 T   h. 22 lb   i. 200 lb

3.

Pounds	1/2	1	2	2 1/2	3	4	5
Ounces	8	16	32	40	48	64	80

Tons	2	3	4	5	10	12	20
Pounds	4,000	6,000	8,000	10,000	20,000	24,000	40,000

4. a. 32 oz; 48 oz; 64 oz   b. 17 oz; 85 oz; 59 oz   c. 68 oz; 45 oz; 83 oz

5. The second baby was 3 oz heavier.

6. 4 apples in 1 lb; 20 apples in 5 lbs.

7. 2 lbs of rice cost $2.88.

8. She put 13 oz of cocoa powder in the bag.

9. a. 1 lb 1 oz; 1 lb 3 oz; 1 lb 7 oz   b. 2 lb; 2 lb 3 oz; 2 lb 14 oz   c. 3 lb 3 oz; 3 lb 7 oz; 3 lb 12 oz

10. a. 8 oz or 1/2 lb   b. 12 oz or 3/4 pound   c. 2 lb 8 oz or 2 1/2 lb
    d. 4 oz or 1/4 lb   e. 5 lb 12 oz or 5 3/4 lb   f. 4 lb 4 oz or 4 1/4 lbs

11. a. 8 oz; 24 oz   b. 4 oz; 36 oz   c. 12 oz; 28 oz

12. a. 1 lb 15 oz   b. 6 lb 11 oz   c. 11 lb 1 oz   d. 8 lb 4 oz

13. a. 3 lb 2 oz   b. 1/4 lb of bread is 4 oz, so Jessie ate more. He ate 3 oz more.
    c. 3 lb 5 oz   d. 4 bags weigh 5 lb; 10 bags weigh 12 lb 8 oz

Metric Units of Weight, p. 187

1. a. 30 kg   b. 2 kg   c. 100 g   d. 20 kg   e. 5 g   f. 50 kg

2.

kilograms	1/2	2	3	3 1/2	5	10	12
grams	500	2,000	3,000	3,500	5,000	10,000	12,000

kilograms	1/2	1	4	4 1/2	6	10	40
grams	500	1,000	4,000	4,500	6,000	10,000	40,000

3. a. 2,000 g; 3,000 g; 4,000 g   b. 1,600 g; 8,080 g; 2,450 g   c. 8,600 g; 5,008 g; 7,041

4. a. 6 kg; 6 kg 700 g; 5 kg 300 g   b. 1 kg 200 g; 6 kg 70 g; 4 kg 770 g

5. a. 3 kg 300 g   b. 6 kg 400 g   c. 10 kg

6. About 7 apples.

7. 6 workbooks.

8. 5 × 400 g is 2000 g , which is 2 kg, and 1 kg + 1 kg is 2 kg, so both
   quantities of chocolate weigh the same.

9. a. 7 kg 520 g   b. 12 kg   c. 7 kg 810 g

10. a. 5 kg 900 g; 2 kg; 9 kg 200 g   b. 1 kg 200 g; 2 kg; 1 kg 100 g

11. 7 kg 350 g total weight.

12. 5 kg 800 in total.

13. You need 950 grams more.

14. You need 5 bags. Total cost: $8.45.

Customary Units of Volume, p. 190

1. a. - e. Check student's answers.

f.

Cups	Ounces
1/2	4
1	8
1 1/2	12
2	16
3	24

2. a.   b.   c.

Quarts	Cups
1/2	2
1	4
2	8
3	12
4	16
5	20

Quarts	Pints
1/4	1/2
1/2	1
1	2
2	4
3	6
4	8

Gallons	Quarts
1/2	2
1	4
2	8
3	12
4	16
5	20

Customary Units of Volume, continued

3. a. 1 quart   b. 4 cups   c. 2 cups   d. 3 pints   e. 1 pint
    f. 2 pints   g. 1 quart   h. 4 cups   i. 1 gallon   j. 2 quarts

4. a. 2 C; 1 C   b. 4 C; 1 C   c. 4 qt; 12 qt   d. 4 pt; 12 C

5. a. 16 oz
   b. The 20-ounce bottle.
   c. The bucket held 3 C.
   d. 2 servings; 6 servings; 4 servings.
   e. A 32-ounce jumbo drink is more.
   f. It is 80 quarts. You need to fill and empty the bucket 8 times.
   g. 20 times
   h. 1 quart is left.

Metric Units of Volume, p. 193

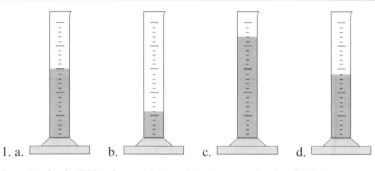

1. a.   b.   c.   d.

2. a. 5 ml   b. 750 ml   c. 10 L   d. 1 L   e. 200 ml   f. 80 L

3.

L	1/2	1	1 1/2	2	5	12
ml	500	1,000	1,500	2,000	5,000	12,000

L	2 1/2	3	4 1/2	8	10	20
ml	2,500	3,000	4,500	8,000	10,000	20,000

4. a. 2,000 ml; 6,000 ml
   b. 1,200 ml; 4,230 ml
   c. 7,070 ml; 4,330 ml
   d. 3 L; 10 L
   e. 4 L 300 ml; 9 L 880 ml
   f. 3 L 40 ml; 5 L 53 ml

5. a. 750 ml   b. 1 L 200 ml   c. 5 glasses; 25 glasses   d. $4.80

6. a. 5 L 150 ml   b. 7 L 700 ml   c. 3 L 50 ml

7. a. 4 L 400 ml; 9 L 200 ml   b. 4 L 400 ml; 1 L 700 ml

8. 1 1/2 L + 400 ml + 200 ml = 1,500 ml + 400 ml + 200 ml = 2,100 ml = 2 L 100 ml

9. 5 × 250 ml + 2 × 2 L + 3 × 350 ml = 1,250 ml + 4 L + 1,050 ml = 6,300 ml = 6 L 300 ml

10. You will buy 7 containers, which will cost $5.46 in total.

Mixed Review, p. 196

1. $24 \times 36 = 20 \times 30 + 20 \times 6 + 4 \times 30 + 4 \times 6 = 600 + 120 + 120 + 24 = 864$ sq units.

2. a. Estimation: $60 \times 30 = 1,800$; Answer: 1,798   b. Estimation: $400 \times 8 = 3,200$; Answer: 3,320
   c. Estimation: $57 \times 100 = 5,700$; Answer: 5,643   d. Estimation: $7 \times 700 = 4,900$; Answer: 4,669

3. a. $3,120   b. 240 km

4. a. 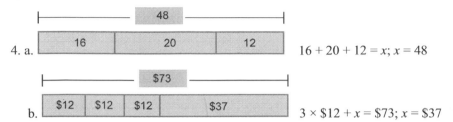 $16 + 20 + 12 = x$; $x = 48$

   b. $3 \times \$12 + x = \$73$; $x = \$37$

5. a. They sold the most strawberries during week 26. About $4,500.
   b. They sold the least strawberries during week 23. The sales were $1,500.
   c. About $12,000.

Review, p. 198

1. a. 6 hours and 52 minutes   b. 10 hours and 40 minutes

2. The plane lands at 6:10 p.m.

3. a. A cold winter day.   b. A nice temperature for indoors.

4. Check the students' work.

   a. 2 3/8 in. ▬▬▬▬▬▬▬▬▬▬

   b. 36 mm ▬▬▬▬▬▬

5. a. 150 mm; 68 mm   b. 12 ft; 68 in   c. 425 cm; 8,000 m

6. 16 ft 10 in.

7. 5 km 600 m total per week

8. a. 16 kg or 34 lb   b. 2 kg   c. 2 oz

9. a. 112 oz, 91 oz   b. 6,200 lb, 7,500 g   c. 2,500 g, 3 kg 456 g

10. He weighed 20 kg 850 g before.

11. The cat food will last five days with 2 ounces left over.

12. a. 3 gal.   b. 3 cups   c. 1/2 gal.

13. a. 2,300 ml, 6 L 550 ml   b. 6 pt, 12 cups   c. 16 qt, 16 fl oz

14. There are 6 six-ounce servings and 4 ounces left over.

15. She paid $60.80.

Math Mammoth Grade 4-B
Answer Key

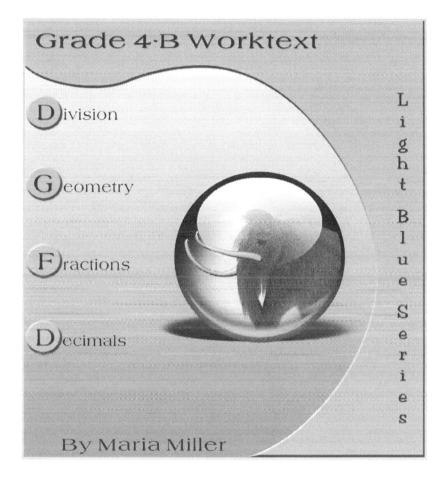

By Maria Miller

Math Mammoth Grade 4-B Answer Key

Contents

Chapter 5: Division

Review of Division, p. 10

1. a. $3 \times 4 = 12$; $12 \div 3 = 4$; $12 \div 4 = 3$
   b. $5 \times 3 = 15$; $15 \div 5 = 3$; $15 \div 3 = 5$
   c. $2 \times 4 = 8$; $8 \div 2 = 4$; $8 \div 4 = 2$

2. a. $21 \div 7 = 3$; $21 \div 3 = 7$; $7 \times 3 = 21$; $3 \times 7 = 21$
   b. $24 \div 4 = 6$; $24 \div 6 = 4$; $4 \times 6 = 24$; $6 \times 4 = 24$
   c. $36 \div 4 = 9$; $36 \div 9 = 4$; $9 \times 4 = 36$; $4 \times 9 = 36$

3. a. 8, 9, 10, $22 \div 2 = 11$, $24 \div 2 = 12$, $26 \div 2 = 13$
   b. 9, 8, 7, $30 \div 5 = 6$, $25 \div 5 = 5$, $20 \div 5 = 4$
   c. 9, 10, 11, $120 \div 10 = 12$, $130 \div 10 = 13$, $140 \div 10 = 14$
   d. 8, 7, 6, $35 \div 7 = 5$, $28 \div 7 = 4$, $21 \div 7 = 3$

4.

Eggs	6	12	24	36	42	54	66	78
Omelets	1	2	4	6	7	9	11	13

Thumbtacks	8	24	32	48	64	80	96	104
Pictures	1	3	4	6	8	10	12	13

5. b. $\$45 - \$34 = \$11$, Jim needs \$11 more.
   c. $400 \div 4 = 100$; each box has 100 apples.
   d. $24 \div 6 = 4$; each person got 4 pieces.
   e. $5 \times 50 = 250$ total books.
   f. $2 \times \$13 = \$26$; Mom paid \$26 for both books.
   g. $20 \div 4 = 5$; there are 5 cows.
   h. $60 \div 3 = 20$; 20 books are on each shelf.

6. a. 9, 10, 5   b. 6, 6, 8   c. 4, 8, 8   d. 8, 3, 5

7. b. $x = 5$ c. $x = 45$ d. $x = 54$

8. a. $10 \times 3 = N$ OR $N = 10 \times 3$; $N = 30$
   b. $9 \times 4 = x$ OR $x = 9 \times 4$; $x = 36$
   c. $20 \times T = 60$, OR $60 = 20 \times T$; $T = 3$
   d. $9 \times y = 81$ OR $81 = y \times 9$; $Y = 9$

9. a. $21 \div 3 = 7$ OR $7 \times 3 = 21$; you can buy 7 books.
   b. $100 \div 5 = 20$ OR $20 \times 5 = 100$; there were 20 apples in each box.
   c. $\$30 \div 5 = \$6$ OR $5 \times \$6 = 30$; each box costs \$6.
   d. $8 \times 5 = 40$; the chocolate bar has 40 squares.
   e. $45 \div 5 = 9$ OR $9 \times 5 = 45$; there are nine fives in 45.
   f. $5 \times 12 = 60$; the boxes weigh 60 pounds.

Division Terms and Division with Zero, p. 13

1. a. 2, the divisor is missing.
   b. 35, the dividend is missing.
   c. 12, the quotient is missing.

2. a. $x \div 7 = 3$; $x = 21$
   b. $140 \div y = 7$; $y = 20$
   c. $150 \div 5 = z$; $z = 30$

3. Answers vary:
   a. $24 \div 4 = 6$, $30 \div 5 = 6$, $60 \div 10 = 6$
   b. $24 \div 2 = 12$, $24 \div 3 = 8$, $24 \div 6 = 4$

4.

Numbers	Product (written)	Product (solved)	Quotient (written)	Quotient (solved)
12 and 3	12×3	36	$36 \div 3$	12
10 and 5	10×5	50	$50 \div 5$	10
20 and 4	20×4	80	$80 \div 4$	20
100 and 10	100×10	1,000	$1,000 \div 10$	100

5. a. 8, 0, 1   b. 11, xx, 1   c. 50, 0, xx   d. 0, 1, xx

6. a. $x = 64$ b. $T = 1$ c. there are many solutions. In fact, x can be any number except 0.   d. $y = 18$

7. Answers vary. Examples:
   a. $24 \div 24 = 1$, $4 \div 4 = 1$ b. $0 \div 36 = 0$, $0 \div 12 = 0$

Puzzle corner. The dividend and quotient both were zeros. For example, he could have had the problems $0 \div 6 = 0$ and $0 \div 9 = 0$.

1.

a. $300 \times 7 = 2,100$	b. $50 \times 800 = 40,000$	c. $60 \times 40 = 2400$
$2100 \div 7 = 300$	$40000 \div 50 = 800$	$2400 \div 60 = 40$
$2100 \div 300 = 7$	$40000 \div 800 = 50$	$2400 \div 40 = 60$

2. a. 50, 5, 5, 50   b. 1,000, 100, 10, 10   c. 6, 60, 6, 60

3. a. 90, 90   b. 900, 900   c. 70, 70

4. a. 40, 4, 400   b. 9, 90, 90   c. 60, 60, 6000

Finding half...	...is the same as dividing by 2!
$\frac{1}{2}$ of 280 is <u>140</u>	$280 \div 2 = $ <u>140</u>

5. a. 40   b. 12,000   c. 330   d. 2,100

6. Dad's paycheck was: $806 + $806 = $1,612.

1/2	
	$806

7. The fisherman had: 1/2 × 800 kg – 350 kg = 50 kg left.

800 kg		
400 kg	350 kg	

half

8.

a. $352 \div 5$	b. $198 \div 4$	c. $403 \div 8$
$\approx 350 \div 5 = 70$	$\approx 200 \div 4 = 50$	$\approx 400 \div 8 = 50$

9.

a. $802 \div 21$	b. $356 \div 61$	c. $596 \div 32$
$\approx 800 \div 20 = 40$	$\approx 360 \div 60 = 6$	$\approx 600 \div 30 = 20$

10. a. $y = 8,000$ b. $s = 4,200$ c. $w = 30$

11.

a. $\approx 80 \div 20 = 4$	b. $\approx 45 \div 5 = 9$
$\approx 120 \div 60 = 2$	$\approx 16,000 \div 400 = 40$
$\approx 2,000 \div 500 = 4$	$\approx 300 \div 30 = 10$

1.

a.	b.	c.	d.
$10 \div 5 = 2$	$9 \div 3 = 3$	$16 \div 2 = 8$	$15 \div 3 = 5$
$\frac{1}{5}$ of 10 is 2.	$\frac{1}{3}$ of 9 is 3.	$\frac{1}{2}$ of 16 is 8.	$\frac{1}{3}$ of 15 is 5.

2.

a. $30 \div 5 = 6$	b. $48 \div 6 = 8$	c. $25 \div 5 = 5$	d. $50 \div 5 = 10$
$\frac{1}{5}$ of 30 is 6.	$\frac{1}{6}$ of 48 is 8.	$\frac{1}{5}$ of 25 is 5.	$\frac{1}{5}$ of 50 is 10.

3.

a. $\frac{1}{6}$ of 30 is 5.	b. $\frac{1}{7}$ of 49 is 7.	c. $\frac{1}{10}$ of 250 is 25.
$30 \div 6 = 5$	$49 \div 7 = 7$	$250 \div 10 = 25$
d. $\frac{1}{2}$ of 480 is 240.	e. $\frac{1}{9}$ of 1,800 is 200.	f. $\frac{1}{5}$ of 400 is 80.
$480 \div 2 = 240$	$1,800 \div 9 = 200$	$400 \div 5 = 80$

4.

a.	b.	c.
$\frac{1}{3}$ of 9 apples is _3_ apples.	$\frac{1}{4}$ of 12 flowers is _3_.	$\frac{1}{5}$ of _15_ fish is _3_ fish.
$\frac{2}{3}$ of 9 apples is _6_ apples.	$\frac{2}{4}$ of 12 flowers is _6_.	$\frac{2}{5}$ of _15_ fish is _6_ fish.
$\frac{3}{3}$ of 9 apples is _9_ apples.	$\frac{3}{4}$ of 12 flowers is _9_.	$\frac{3}{5}$ of _15_ fish is _9_ fish.
	$\frac{4}{4}$ of 12 flowers is _12_.	$\frac{4}{5}$ of _15_ fish is _12_ fish.
		$\frac{5}{5}$ of _15_ fish is _15_ fish.

5. a. 4, 8, 12   b. 4, 12, 20   c. 50, 150, 350
   d. 70, 140, 350   e. 30, 210, 330   f. 7, 63, 350

6. a. Marsha got $18 from her mom. She put into her savings $6, which was one-third part of it. $18 \div 6 = 3$
   b. Mariana spent one-fourth of her $80 savings, or _$20_. $80 \div 4 = 20$
   c. One-fifth (which was 5 boys) of all the boys went jogging. So, in total there were _25_ boys. $25 \div 5 = 5$

7. a. One pound is one-eighth part of the bag. It costs $0.40.
   b. Five-eighths costs $2.00.

8. a. One-tenth of the pie weighs 120 grams.
   b. Nine-tenths of the pie weighs 1,080 grams.

9. $28 \div 4 = 7$, and $7 \times 3 = 21$. Three would cost $21.

10. Mark: $12.20   Judy: $6.10   Art: $3.05   Grace: $3.05

11. Erica and James each had 28 balloons to sell. By the evening, Erica had sold 1/2 or 14 of her balloons.
    James had sold 3/4 or 21 of his. Together they had sold 35 balloons. They took in $35 \times 1.20 = \$42$.

Order of Operations and Division, p. 20

1. a. 3   b. 100   c. 120   d. 2,000

2. a. 62   b. 152   c. 2,000   d. 18

3. a. 9   b. 17   c. 200   d. 5

4.

a.	b.	c.
$24 \div 2 + 10 = 22$	$18 + 30 \div 2 = 33$	$40 - 40 \div 8 = 35$
$24 \div (2 + 10) = 2$	$(18 + 30) \div 2 = 24$	$(40 - 40) \div 8 = 0$

5. a. $(20 + 15) \div 5 = 7$
   b. $20 - 50 \div 5 = 10$
   c. $20 \times 30 - 100 = 500$

6. $(21 + 17) \div 2$. The answer is 19 figures.

7. $6 \times 6 \div 4$. The answer is $9.

8. a. 5; 7   b. 60; 120   c. 20; 20   d. 1; 1   e. 0; 0

9. a. $5 \div 5 \times 5 = 5$
   b. $(5 - 5) \times 5 = 0$
   c. $(5 + 5) \div 5 = 2$
   d. $(5 + 5) \times (5 + 5) = 100$
   e. $5 \times 5 + 5 - 5 = 25$

Puzzle corner:
$(5 - 5) \times 5 = 0$
$5 \div 5 = 1$
$(5 + 5) \div 5 = 2$
$(5 + 5 + 5) \div 5 = 3$
$(5 \times 5 - 5) \div 5 = 4$
$5 \times 5 \div 5 = 5$
$(5 \times 5 + 5) \div 5 = 6$
$(5 \times 5 + 5 + 5) \div 5 = 7$
$(5 + 5) - (5 + 5) \div 5 = 8$
$(5 + 5) - (5 \div 5) = 9$
$5 + 5 = 10$

The Remainder, Part 1, p. 22

1. a. $10 \div 3 = 3 R1$ b. $17 \div 5 = 3 R2$ c. $11 \div 4 = 2 R3$

2. a. $10 \div 3 = 3 R1$ b. $8 \div 3 = 2 R2$ c. $19 \div 5 = 3 R4$
   d. $15 \div 4 = 3 R3$ e. $18 \div 4 = 4 R2$ f. $9 \div 2 = 4 R1$

3.

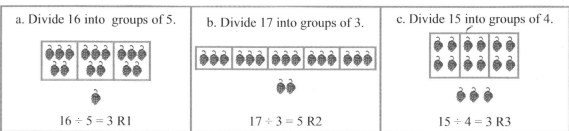

a. $17 \div 4 = 4 R1$	b. $9 \div 2 = 4 R1$	c. $11 \div 6 = 1 R5$

4. a. $10 \div 4 = 2 R2$ b. $17 \div 2 = 8 R1$ c. $12 \div 5 = 2 R2$

5.

a. Divide 16 into groups of 5.	b. Divide 17 into groups of 3.	c. Divide 15 into groups of 4.
$16 \div 5 = 3 R1$	$17 \div 3 = 5 R2$	$15 \div 4 = 3 R3$

6.

a. $27 \div 5 = 5 R2$ 5 goes into 27 five times.	b. $16 \div 6 = 2 R4$ 6 goes into 16 two times.	c. $11 \div 2 = 5 R1$ 2 goes into 11 five times.
d. $37 \div 5 = 7 R2$	e. $26 \div 3 = 8 R2$	f. $56 \div 9 = 6 R2$
g. $43 \div 5 = 8 R3$	h. $34 \div 6 = 5 R4$	i. $40 \div 7 = 5 R5$

The Remainder, Part 1, continued

7.

a.	b.	c.
$23 \div 4 = 5$ R3	$16 \div 7 = 2$ R2	$21 \div 8 = 2$ R5
$23 \div 5 = 4$ R3	$20 \div 3 = 6$ R2	$12 \div 9 = 1$ R3

8.

a. $10 \div 5 = 2$ R 0	b. $17 \div 3 = 5$ R2	c. $12 \div 4 = 3$ R0
$11 \div 5 = 2$ R1	$18 \div 3 = 6$ R0	$13 \div 4 = 3$ R1
$12 \div 5 = 2$ R2	$19 \div 3 = 6$ R1	$14 \div 4 = 3$ R2
$13 \div 5 = 2$ R3	$20 \div 3 = 6$ R2	$15 \div 4 = 3$ R3
$14 \div 5 = 2$ R4	$21 \div 3 = 7$ R0	$16 \div 4 = 4$ R0
$15 \div 5 = 3$ R0	$22 \div 3 = 7$ R1	$17 \div 4 = 4$ R1

9. a. $27 \div 5 = 5$ R2
   b. $19 \div 5 = 3$ R4
   c. $36 - 3 \div 6 = 5$ R3
   d. No, because $51 \div 8 = 6$ R3.
   e. Of four, no. $35 \div 4 = 8$ R3 (the division is not even.) Of five, yes. $35 \div 5 = 7$.
      Of six, no. $35 \div 6 = 5$ R5. Of seven, yes. $35 \div 7 = 5$.
   f. $38 \div 6 = 6$ R2. There were two photos on the last page. Six pages were full.

The Remainder, Part 2, p. 25

1. a. 3   b. 9   c. 7   d. 9

2. a.
$$\begin{array}{r} 6 \\ 5)\overline{3\,2} \\ -3\,0 \\ \hline 2 \end{array}$$
b.
$$\begin{array}{r} 8 \\ 5)\overline{4\,4} \\ -4\,0 \\ \hline 4 \end{array}$$
c.
$$\begin{array}{r} 6 \\ 6)\overline{3\,7} \\ -3\,6 \\ \hline 1 \end{array}$$
d.
$$\begin{array}{r} 4 \\ 7)\overline{2\,9} \\ -2\,8 \\ \hline 1 \end{array}$$

e.
$$\begin{array}{r} 5 \\ 8)\overline{4\,6} \\ -4\,0 \\ \hline 6 \end{array}$$
f.
$$\begin{array}{r} 5 \\ 9)\overline{5\,2} \\ -4\,5 \\ \hline 7 \end{array}$$
g.
$$\begin{array}{r} 8 \\ 4)\overline{3\,5} \\ -3\,2 \\ \hline 3 \end{array}$$
h.
$$\begin{array}{r} 6 \\ 9)\overline{5\,7} \\ -5\,4 \\ \hline 3 \end{array}$$

3. a. $6 \times 5 + 2 = 32$ b. $8 \times 5 + 4 = 44$ c. $6 \times 6 + 1 = 37$ d. $4 \times 7 + 1 = 29$
   e. $5 \times 8 + 6 = 46$ f. $5 \times 9 + 7 = 52$ g. $8 \times 4 + 3 = 35$ h. $6 \times 9 + 3 = 57$

4. $33 \div 6 = 5$ R3. Jill needed six containers, but only five were full.

5. $100 \div 42 = 2$ R16. They needed three buses to haul the children.

6. There were two teams of seven and one team of six players.

7. $73 \div 20 = 3$ R13. She needed four folders. Three folders were full.

8. $3 \times 23 + 15 = 84$. He had 84 award stickers.

9. $36 \div 11 = 3$ R3. She put three pencils back into the cabinet.

10.

| a. $12 \div 3 = 4$ R0 | b. $10 \div 2 = 5$ R0 | c. $19 \div 4 = 4$ R3 |

The Remainder, Part 2, continued

11.

a. 21 ÷ 5 = 4 R1	b. 56 ÷ 8 = 7 R0	c. 43 ÷ 7 = 6 R1
22 ÷ 5 = 4 R2	57 ÷ 8 = 7 R1	44 ÷ 7 = 6 R2
23 ÷ 5 = 4 R3	58 ÷ 8 = 7 R2	45 ÷ 7 = 6 R3
24 ÷ 5 = 4 R4	59 ÷ 8 = 7 R3	46 ÷ 7 = 6 R4

12. The shortcut is: the remainder is always the last digit of the dividend (the number you divide), and the other digits are the quotient (the answer)

a. 29 ÷ 10 = 2 R9	b. 78 ÷ 10 = 7 R8	c. 54 ÷ 10 = 5 R4
30 ÷ 10 = 3 R0	79 ÷ 10 = 7 R9	55 ÷ 10 = 5 R5
31 ÷ 10 = 3 R1	80 ÷ 10 = 8 R0	56 ÷ 10 = 5 R6

Puzzle Corner:
a. 16 ÷ 5 = 3 R1 OR 16 ÷ 3 = 5 R1
b. 31 ÷ 7 = 4 R3 OR 31 ÷ 4 = 7 R3
c. 135 ÷ 4 = 30 R3 OR 135 ÷ 30 = 4 R3

Long Division 1, p. 28

1.

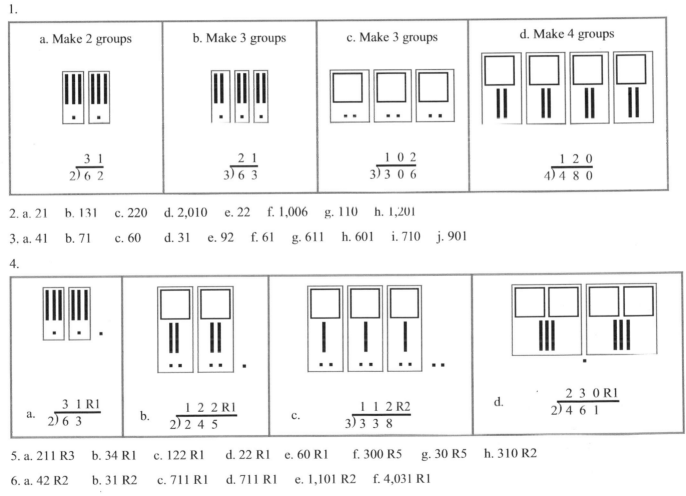

2. a. 21   b. 131   c. 220   d. 2,010   e. 22   f. 1,006   g. 110   h. 1,201

3. a. 41   b. 71   c. 60   d. 31   e. 92   f. 61   g. 611   h. 601   i. 710   j. 901

4.

5. a. 211 R3   b. 34 R1   c. 122 R1   d. 22 R1   e. 60 R1   f. 300 R5   g. 30 R5   h. 310 R2

6. a. 42 R2   b. 31 R2   c. 711 R1   d. 711 R1   e. 1,101 R2   f. 4,031 R1

7. a. 110, 410   b. 9, 123   c. 412, 6 R20

Long Division 2, p. 32

1. a. 16   b. 24   c. 29   d. 15   e. 19   f. 39   g. 26   h. 47

2. a. 47   b. 43   c. 64   d. 34   e. 84   f. 58

Long Division 3, p. 35

1. a. 115   b. 123   c. 244   d. 276   e. 318   f. 121
   g. 113   h. 113   i. 325   j. 113   k. 112   l. 218

2. a. 189   b. 166   c. 142   d. 117   e. 152   f. 117

Long Division with 4-Digit Numbers, p. 39

1. a. 2,347   b. 2,310   c. 1,785   d. 4,885

2. a. 1,934   b. 551   c. 1,340
   d. 1,038   e. 1,317   f. 1,216

3. a. 493   b. 384   c. 924   d. 49   e. 87   f. 371

4. 9 × $16 ÷ 2 = $72. They each paid $72.

5. 1,092 min ÷ 7 = 156 min or 2 h 36 min each day

6. 2600 ÷ 8 = 325. The second clue is at 325 feet.
   The third clue is at 650 feet.

7. 96 ÷ 6 = 16. Sixteen children were coming to the party.
   8 × 25 = 200 and 200 − 96 = 104. She had 104 balloons left.

More Long Division, p. 43

1. a. 1,045   b. 1,406   c. 2,037   d. 1,307

2. a. 2,705   b. 1,308   c. 1,309   d. 1,063

3. a. 108   b. 205   c. 402   d. 405   e. 308   f. 1,070

4. a. 285 ÷ 5 = 57.
   There are 57 buttons in one compartment.
   b. 3 × 57 = 171.
   There are 171 buttons in three compartments.

4. c. 4 × 57 = 228.
   There are 228 buttons in four compartments.

5. a. $12.96 ÷ 8 = $1.62.
   You will pay $1.62. Your brother will pay $1.62.
   Mom will pay $12.96 − $1.62 − $1.62 = $9.72.
   b. You get two cups. Your brother gets two cups.
   Mom gets 12 cups.

6. a. 21,234   b. 35,407   c. 21,645   d. 3,162   e. 5,275

Remainder Problems, p. 46

1. a. 171 R1   Check: 3 × 171 + 1 = 514
   b. 84 R1   Check: 8 × 84 + 1 = 673
   c. 317 R3   Check: 6 × 317 + 3 = 1,905
   d. 2,051 R1 Check: 4 × 2,051 + 1 = 8,205

2. a. wrong; 77 R1   b. right   c. wrong: 451
   d. The remainder is larger than the divisor.

3. a. 112 ÷ 9 = 12 R4. We get 12 rows, 9 chairs each row,
   and 4 chairs will be left over or put in an extra row.
   b. 800 ÷ 3 = 266 R2. We get 266 piles, 3 erasers in each
   pile, and 2 erasers left over.

4. They will get 166 full sacks.

5. 20 × 50 = 1,000 and 19 × 50 = 950. So, 19 buses is
   enough to transport 950 people.

6. 75 ÷ 4 = 18 R3. One 18-day vacation and three 19-day
   vacations. If the division had been even, all of the
   vacations would have been 18 days, but now there are
   three extra days to be added to three of the vacations.

7. 400 − (2 × 90) − (4 × 40) = 60; 60 ÷ 6 = 10.
   They will have ten full 6-kg boxes of strawberries.

8. a. Yes. There will be 103 containers. 412 ÷ 4 = 103.
   b. No, there will be 82 containers with 2 left over.
   412 ÷ 5 = 82 R2.
   c. No, there will be 68 with four left over.
   412 ÷ 6 = 68 R4.

9. 740 ÷ 6 = 123 R2. Paint 123 blocks in four of the colors
   (any four), and 124 blocks in the two remaining colors.

10. a. 70 R1; 70 R2; 70 R3   b. 172 R2, 172 R3, 172 R4
    c. 82 R1; 82 R2; 82 R3   d. 798 R3   798 R4   798 R5
    You can figure out the two other problems after solving
    one, because the remainder will increase by one as
    the dividend increases by one.

11. It would be 38 R4. The only difference is that the
    remainder increases by 1.

12. a. 78 R7; 6 R6; 34   b. 45 R2; 50 R9; 5 R2
    c. 46 R3; 98 R2; 92 R5
    The ones digit of the dividend will always be the
    remainder.

Puzzle Corner: The remainder is larger than the divisor.

Long Division with Money, p. 50

1. a. $8.47   b. $3.72

2. a. $28.50   b. $1.14

3. $25.56 + $3.55 + $2.75 = $31.86
   $31.86 ÷ 2 = $15.93. Each girl paid $15.93.

4. ($25.95 + $4.73) ÷ 3 = $10.10. Each person's share
   was $10.10

5. Her bill was $3.52 ÷ 4 × 3 = $2.64.

6. $358.60 − $100 = $258.60; $258.60 ÷ 4 = $64.65.
   Each payment was $64.65.

Long Division Crossword Puzzle, p. 52

1. Across:
   a. 3,440 ÷ 8 = 430
   b. 574 ÷ 7 = 82
   c. 234 ÷ 9 = 26
   d. 1,707 ÷ 3 = 569
   e. 4,756 ÷ 2 = 2,378

Down:
   a. 1,072 ÷ 8 – 134
   b. 6,135 ÷ 3 = 2,045
   c. 145 ÷ 5 = 29
   d. 2,652 ÷ 4 = 663
   e. 1,442 ÷ 7 = 206
   f. 3,474 ÷ 9 = 386

a. 1				
3		**b.** 2	**b.** 8	**e.** 2
a. 4	3	0		0
		4	**c.** 2	6
	d. 5	**d.** 6	9	
		6		**f.** 3
	e. 2	3	7	8
				6

Average, p. 53

1. (78 + 87 + 69 + 86) ÷ 4 = 80. Judith's average score is 80.

2. (18 + 22 + 26 + 23 + 16) ÷ 5 = 21. The average temperature for the day was 21°C.

3. 414 ÷ 6 = 69. Dad averaged 69 km in one hour.

4. 12 × 55 = 660. A dozen eggs would weigh 660 grams.

5. 7 × 76 = 532. It cost $532.

6. (234 + 178 + 250 + 198) ÷ 4 = 215. Her weekly average grocery bill was $215.

7. The girls' average time was 15 minutes. The boys' average time was 13 minutes. The boys are faster.
   The difference is two minutes.

8. a.

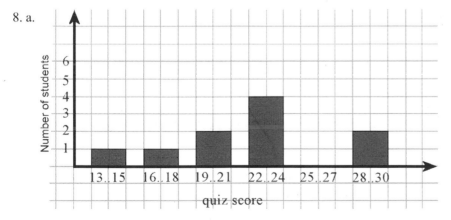

Quiz score	Frequency
13..15	1
16..18	1
19..21	2
22..24	4
25..27	0
28..30	2

b. The average score is 22.
c. Look at the "peak" of the graph. The average is usually near that point.

9. a. The average age is 29.   b. Now the average age is 34.

Puzzle corner: 213 ÷ 12 is 17 R9.

Problems with Fractional Parts, p. 56

1. a. One slice weighs 20 grams.
   b. Three slices weigh 60 grams.
   c. Eleven slices weigh 220 grams.

2. a. Two-sixths of the pie weighs 150 grams.
   b. It weighs 375 grams.

3. If you need to calculate 5/9 of the number 729, first divide 729 by 9, then multiply the result by 5.
   5/9 of 729 is 405

4. $36.50 ÷ 5 × 2 = $14.60

5. The other washer costs $452 ÷ 4 × 3 = $339.

6. 12,600 ÷ 9 × 2 = 2,800
   a. 9,800 miles left   b. 2,800 miles

7. $268 ÷ 4 × 3 = $201. It cost $201. She has $67 left.

8. a. 6 tons, or 12,000 pounds.

9. Edward worked (56 ÷ 4) × 3 = 42 hours.
   James worked 21 hours.
   Together they worked 56 + 42 + 21 = 119 hours.

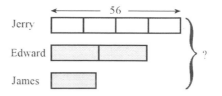

Problems to Solve, p. 58

1. $3.25 + 3 × $3.25 = $13

2. 1,200 are females. Since there are three times as many females as males, we can divide the 1,600 workers into four parts. One-fourth part of 1,600 is 400. So, there are 3 × 400 or 1,200 female workers.

3. Cindy has $14 left. (Half of Cindy's money is $14.)

4. 96 workers. Since 84 is 7/8 of the workers, 84 ÷ 7 = 12 gives you 1/8 of the workers, and then 8 × 12 = 96 is the total amount.

5. Mary got 16 pieces. One-third of the pieces is 8 pieces.

6. 2 yards, 2 feet, 6 inches.
   One-fifth of 15 yards is 3 yards. Now subtract
   3 yards − 6 inches = 2 yards 2 feet 6 inches.

7. There are 6 speckled chickens. Solution: there are 18 white chickens (three that were sold and 15 that were left), and that is 3/4 of all the chickens.

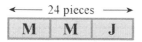

So 18 ÷ 3 = 6 gives you 1/4 of the chickens (one block). And then, one block, or six chickens, are speckled.

8. They would both cost $9 after the discount.
   One-fourth of $12 is $3, so the new price is $9.
   One-third of $13.50 is $4.50, so the new price is $9.

9. Jackie paid $5.00. First find the total without the discount:
   5 × $1.50 = $7.50. One-third of that is $2.50. The price with discount is then $7.50 − $2.50 = $5.00

Divisibility, p. 61

1. a. 7; yes   b. 6 R4; no   c. 3 R2; no   d. 12; yes

2. a. 24 R2, no   b. 86 R1; no   c. 418 R2; no

3. Here is a multiplication fact: 8 × 9 = 72. So, 8 is a <u>factor</u> of 72, and so is 9.
   Also, 72 is a <u>multiple</u> of 8, and also 72 is a <u>multiple</u> of 9.
   And, 72 is <u>divisible</u> by 8 and also by 9.

Divisibility, continued

4

a. Is 5 a factor of 55? Yes, because $5 \times 11 = 55$.	b. Is 8 a divisor of 45? No, because $45 \div 8 = 5$ R5.
c. Is 36 a multiple of 6? Yes, because $6 \times 6 = 36$.	d. Is 34 a multiple of 7? No, because $34 \div 7 = 4$ R6.
e. Is 7 a factor of 46? No, because $46 \div 7 = 6$R4. (It is not an even division.)	f. Is 63 a multiple of 9? Yes, because $7 \times 9 = 63$.

5. a. 0, 11, 22, 33, 44, 55, 66, 77, 88, 99, 110, 121, 132, 143, 154
   b. 0, 111, 222, 333, 444, 555, 666, 777, 888, 999, 1,110, 1,221, 1,332, 1,443, 1,554, 1,665

6.

number	divisible by 2	divisible by 5	number	divisible by 2	divisible by 5	number	divisible by 2	divisible by 5	number	divisible by 2	divisible by 5
750	x	x	755		x	760	x	x	765		x
751			756	x		761			766	x	
752	x		757			762	x		767		
753			758	x		763			768	x	
754	x		759			764	x		769		

7.

number	divisible by 2	divisible by 5	divisible by 10	number	divisible by 2	divisible by 5	divisible by 10	number	divisible by 2	divisible by 5	divisible by 10
860	x	x	x	865		x		870	x	x	x
861				866	x			871			
862	x			867				872	x		
863				868	x			873			
864	x			869				874	x		

If a number is divisible by 10, it ends in zero, so it is ALSO divisible by __2__ and __5__.

8. a. 2, <u>4</u>, 6, <u>8</u>, 10, <u>12</u>, 14, <u>16</u>, 18, <u>20</u>, 22, <u>24</u>, 26, <u>28</u>, 30, <u>32</u>, 34, <u>36</u>, 38, <u>40</u>, 42, <u>44</u>, 46, <u>48</u>, 50, <u>52</u>, 54, <u>56</u>, 58, <u>60</u>
   This is also a list of multiples of (or multiplication table of) 2.

   b. 2, <u>4</u>, 6, <u>8</u>, 10, <u>12</u>, 14, <u>16</u>, 18, <u>20</u>, 22, <u>24</u>, 26, <u>28</u>, 30, <u>32</u>, 34, <u>36</u>, 38, <u>40</u>, 42, <u>44</u>, 46, <u>48</u>, 50, <u>52</u>, 54, <u>56</u>, 58, <u>60</u>
   These are every other number in the list of multiples of 2.

   c. 2, <u>4</u>, 6, <u>8</u>, 10, <u>12</u>, 14, <u>16</u>, 18, <u>20</u>, 22, <u>24</u>, 26, <u>28</u>, 30, <u>32</u>, 34, <u>36</u>, 38, <u>40</u>, 42, <u>44</u>, 46, <u>48</u>, 50, <u>52</u>, 54, <u>56</u>, 58, <u>60</u>
   These are every third number in the list of multiples of 2, or every third even number divisible by 6.

   d. 12, 24, 36, 48, and 60 - or multiples of 12.

9. a. 3, 6, 9, 12, 15, 18, 21, 24, 27, 30, 33, 36, 39, 42, 45, 48, 51, 54, 57, 60
   This is also a list of multiples of (or multiplication table of) 3.

   b. 3, <u>6</u>, 9, <u>12</u>, 15, <u>18</u>, 21, <u>24</u>, 27, <u>30</u>, 33, <u>36</u>, 39, <u>42</u>, 45, <u>48</u>, 51, <u>54</u>, 57, <u>60</u>
   These are every second number in the list of multiples of 3.

   c. 3, <u>6</u>, 9, <u>12</u>, 15, <u>18</u>, 21, <u>24</u>, 27, <u>30</u>, 33, <u>36</u>, 39, <u>42</u>, 45, <u>48</u>, 51, <u>54</u>, 57, <u>60</u>
   These are every third number in the list of multiples of 3.

10. 18, 36, 54

11. 1

12. It is also a multiple of 1, 2, 10, and 20.

Mystery number: 33 and 60

number	divisible by 1	divisible by 2	divisible by 3	divisible by 4	divisible by 5	divisible by 6	divisible by 7	divisible by 8	divisible by 9	divisible by 10
2	x	x								
3	x		x							
4	x	x		x						
5	x				x					
6	x	x	x			x				
7	x						x			
8	x	x		x				x		
9	x		x						x	
10	x	x			x					x
11	x									
12	x	x	x	x		x				
13	x									
14	x	x					x			
15	x		x		x					
16	x	x		x				x		
17	x									
18	x	x	x			x			x	
19	x									
20	x	x		x	x					x
21	x		x				x			
22	x	x								
23	x									
24	x	x	x	x		x		x		
25	x				x					
26	x	x								
27	x		x						x	
28	x	x		x			x			
29	x									
30	x	x	x		x	x				x
31	x									
32	x	x		x				x		
33	x		x							
34	x	x								
35	x				x		x			

2. Prime numbers: 2, 3, 5, 7, 11, 13, 17, 19, 23, 29, 31

3. Answers vary, as you can write a composite number as a product in many different ways.

a. 33 is composite.
  $33 = 3 \times 11$
	b. 52 is composite.
	  $52 = 2 \times 26$
		c. 41 is prime.

d. 39 is composite.
  $39 = 3 \times 13$
	e. 43 is prime.	f. 45 is composite.
		  $45 = 5 \times 9$

Prime Numbers, continued

1.

number	digit sum	divisible by 3?	number	digit sum	divisible by 3?
98	17	no	888	24	yes
105	6	yes	1,045	10	no
567	18	yes	1,338	15	yes
59	14	no	612	9	yes

5.

number	divisible by 7?	number	divisible by 7?	number	divisible by 7?
99	no	24	no	85	no
74	no	100	no	63	yes
56	yes	84	yes	105	yes

6. Answers vary, as you can write a composite number as a product in many different ways.

a. 67 is prime.	b. 57 is composite. $57 = 3 \times 19$	c. 47 is prime.
d. 53 is prime.	e. 63 is composite. $63 = 7 \times 9$	f. 61 is prime.
g. 93 is composite. $93 = 3 \times 31$	h. 85 is composite. $85 = 5 \times 17$	i. 91 is composite. $91 = 7 \times 13$
j. 87 is composite. $87 = 3 \times 29$	k. 79 is prime.	l. 97 is prime.

Finding Factors, p. 68

1.

a. factors: 1, 2, 3, 6	b. factors: 1, 2, 5, 10
c. factors: 1, 2, 3, 4, 6, 12	d. factors: 1, 3, 5, 15
e. factors: 1, 2, 4, 5, 10, 20	f. factors: 1, 2, 3, 6, 9, 18

2. Only Olivia's work was totally correct.

a. Aiden found all the factors of 34:	b. Olivia found all the factors of 28:
~~$34 = 2 \times 18$~~ $34 = 2 \times 17$ ~~$34 = 1 \times 17$~~ $34 = 1 \times 34$ The factors are 1, 2, 17, ~~18~~, 34	$28 = 1 \times 28$ $28 = 2 \times 14$ $28 = 4 \times 7$ The factors are 1, 2, 4, 7, 14, and 28.
c. Jayden found all the factors of 33:	d. Isabella found all the factors of 36:
$33 = 1 \times 33$ ~~$33 = 3 \times 13$~~ $33 = 3 \times 11$ The factors are 1, 3, ~~13~~, 11, 33.	$36 = 6 \times 6$ ~~$36 = 3 \times 12$~~ $36 = 3 \times 12$ $36 = 4 \times 9$ ~~$36 = 1 \times 36$~~ The factors are 4, 6, and 9. <u>Also 1, 2, 3, 12, 18, 36</u>

Finding Factors, continued

3.

a. factors: 1, 2, 23, 46	b. factors: 1, 2, 4, 17, 34, 68
c. factors: 1, 3, 9, 11, 33, 99	d. factors: 1, 2, 3, 4, 6, 8, 9, 12, 18, 24, 36, 72
e. factors: 1, 73	f. factors: 1, 2, 4, 5, 8, 10, 16, 20, 40, 80
g. factors: 1, 5, 19, 95	h. factors: 1, 2, 4, 8, 16, 32, 64

Mixed Review, p. 70

1. a. 4,284   b. 49,068

2. a. 84; 80   b. 20; 54   c. 1,090; 90

3.

Estimate: $1,568 + 4,839 + 452$ ↓ ↓ ↓ $\approx 1,600 + 4,800 + 500 = 6,900$	Exact: 6,859

4. a. 1,998; 3,960; 3,991   b. 6,990; 9,970; 991   c. 1,900; 6,700; 9,400

5. a. 34,268   b. 800,046   c. 406,780

6.

a. 3 ft = 36 in
    9 ft = 108 in
	b. 2 ft 5 in = 29 in
	    7 ft 8 in = 92 in
		c. 9 ft 2 in = 110 in
		    10 ft 11 in = 131 in

7. a. $5,400 = 90 \times 60$ b. $16 \times 20 = 8 \times 40$

   c. $7 \times 49 + 49 = 8 \times 49$ d. $24,000 = 300 \times 80$

   e. $7 \times 13 = 5 \times 13 + 26$ f. $1,500 - 500 = 5 \times 200$

8. a. Estimate: 8 weeks (8 × $50 = $400). Exact: 9 weeks, because 9 × $45 = $405. He will have $6 left over.
   b. She needs 230 cm of string, 69 sheets of paper, and 46 rolls of tissue paper.

   c. James had 25 marbles.

←————— 100 —————→
Greg	James	Mark

Review, p. 72

1.

a.	b.	c.
$20 \div 10 + 15 = 17$ $20 \times 10 + 15 = 215$	$(200 + 100) \div 5 = 60$ $200 + 100 \div 5 = 220$	$10 \times 12 + 40 \div 10 = 124$ $10 \times (12 + 40) \div 10 = 52$

2.

a.	b.	c.
$3,100 \div 100 = 31$ $450 \div 10 = 45$	$240 \div 20 = 12$ $800 \div 40 = 20$	$4,200 \div 600 = 7$ $3,200 \div 80 = 40$

3.

a.	b.	c.
$45 \div 6 = 7$ R3	$12 \div 7 = 1$ R5	$31 \div 4 = 7$ R3
$46 \div 6 = 7$ R4	$27 \div 8 = 3$ R3	$56 \div 9 = 6$ R2

4. a. 236   b. 188

5. a. 78 R2   b. 474 R1

6. $288 \div 4 = 72$. Timmy has 72 seashells.

7. a. $70 \div 12 = 5$ R10. Mark had five full boxes of candles.
   b. One box had ten candles.

8. $\$38.88 \div 4 = \$9.72$. One yard cost $\$9.72$.

9. $(92 + 85 + 89 + 75 + 89) \div 5 = 86$. John's average score was 86.

10.

Number	13	40	57	135	354	2,380
Divisible by 3			x	x	x	
Divisible by 5		x		x		x
Divisible by 10		x				x

11.

a. Is 7 a factor of 64?
   No, because it does not divide evenly into 64.
   OR No, because $64 \div 7 = 9$ R1; there is a remainder.
	b. Is 98 a multiple of 2?
	   Yes, because it is an even number.
	   OR Yes, because $2 \times 49 = 98$.

c. Is 76 divisible by 8?
   No, because $76 \div 8 = 9$ R4; the division is not even.
	d. Is 30 a factor of 30?
	   Yes, because $1 \times 30 = 30$.

12.

a. 87 is composite.
   $87 = 3 \times 29$
	b. 89 is prime.	c. 91 is composite.
		   $91 = 7 \times 13$

13.

a. factors: 1, 2, 3, 4, 6, 8, 12, 24	b. factors: 1, 3, 9, 27
c. factors: 1, 2, 3, 6, 11, 22, 33, 66	d. factors: 1, 3, 5, 15, 25, 75

Puzzle corner:
5, 11, 17, 23, 29, 35, 41, 47, 53, 59, 65, 71, 77, 83, 89, 95

Chapter 6: Geometry

Review: Area of Rectangles, p. 79

1. a. $3 \text{ cm} \times 2 \text{ cm} = 6 \text{ cm}^2$ b. $1 \text{ ft} \times 1 \text{ ft} = 1 \text{ ft}^2$ c. $4 \text{ km} \times 8 \text{ km} = 32 \text{ km}^2$

2. a. 5 in.   b. 10 cm   c. 25 ft

3.

a. $12 \text{ cm} \times s = 96 \text{ cm}^2$ $s = 8 \text{ cm}$	b. $20 \text{ cm} \times s = 1,000 \text{ cm}^2$ $s = 50 \text{ cm}$

4. a. $9 \text{ m} \times s = 45 \text{ m}^2$ $s = 5 \text{ m}$ b. $60 \text{ ft} \times s = 1,800 \text{ ft}^2$ $s = 30 \text{ ft}$

5. a. $36 \text{ in.} \times 24 \text{ in.} = 864 \text{ in.}^2$ (the student needs to use the multiplication algorithm with pencil and paper)
   b. $3 \text{ ft} \times 2 \text{ ft} = 6 \text{ ft}^2$

6. a. $4 \times (2 + 5) = 4 \times 2 + 4 \times 5 = 8 + 20 = 28$ b. $3 \times (5 + 2) = 3 \times 5 + 3 \times 2 = 15 + 6 = 21$

7. a. $A = 30 \text{ ft} \times (30 \text{ ft} + 75 \text{ ft})$
   $= 30 \text{ ft} \times 30 \text{ ft} + 30 \text{ ft} \times 75 \text{ ft}$
   $= 900 \text{ ft}^2 + 2,250 \text{ ft}^2 = 3,150 \text{ ft}^2$

   b. $2,250 \text{ ft}^2 - 900 \text{ ft}^2 = 1350 \text{ ft}^2$

8. Check students' drawings. a.   2 in.   b. 3 cm   c. 1 ft

9. a. There are several possible number sentences, depending on how you divide the shape into rectangles.
   $4 \times 5 + 3 \times 2 + 2 \times 6 = 20 + 6 + 12 = 38$ square units
   or $4 \times 2 + 8 \times 3 + 2 \times 3 = 8 + 24 + 6 = 38$ square units

   b. Here it is handy to use subtraction. $7 \times 6 - 3 \times 2 = 36$ square units

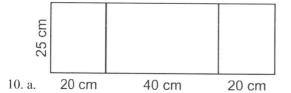

10. a.   20 cm   40 cm   20 cm

   b. $25 \text{ cm} \times (20 \text{ cm} + 40 \text{ cm} + 20 \text{ cm}) = 25 \text{ cm} \times 80 \text{ cm} = 2,000 \text{ cm}^2$.

11. Answers vary. Check student's answers. For example:

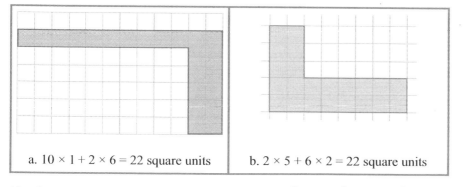

a. $10 \times 1 + 2 \times 6 = 22$ square units   b. $2 \times 5 + 6 \times 2 = 22$ square units

12. The area is $48 \text{ ft} \times 48 \text{ ft} - 30 \text{ ft} \times 20 \text{ ft} = 2304 \text{ ft}^2 - 600 \text{ ft}^2 = 1704 \text{ ft}^2$.

13. a. $3 \text{ m} \times 3\text{m} + 7 \text{ m} \times 5 \text{ m} + 3 \text{ m} \times 3 \text{ m} = 9 \text{ m}^2 + 35 \text{ m}^2 + 9 \text{ m}^2 = 53 \text{ m}^2$.
   b. Divide the shape into two rectangles. That can be done in two ways. One way results in the calculation
   $16 \text{ ft} \times 32 \text{ ft} + 24 \text{ ft} \times 12 \text{ ft} = 512 \text{ ft}^2 + 288 \text{ ft}^2 = 800 \text{ ft}^2$.

Review: Area and Perimeter, p. 84

1. a. perimeter b. volume c. area

2. a. perimeter = 32 in. area = 48 in.^{2} b. perimeter = 200 cm area = 2,500 cm^2

3. a. perimeter = 16 cm area = 7 cm^2
   b. perimeter = 32 cm area = 28 cm^2. Notice that when the perimeter gets doubled, the area gets *quadrupled*.

4. You can of course also use a letter or other symbol for the unknown, instead of ?.

   a. 9 ft + ? + 9 ft + ? = 32 ft or 9 ft + ? = 16 ft. Solution: ? = 7 ft.

   b. 15 cm + ? + 15 cm + ? = 44 cm or 15 cm + ? = 22 cm. Solution: ? = 7 cm.

   c. ? + ? + ? + ? = 24 m or ? + ? = 12 m or 4 × ? = 24 m. Solution: ? = 6 m.

   d. 5 cm. The number sentence would be $s \times s = 25$ cm^2. Its perimeter is 20 cm.

5.

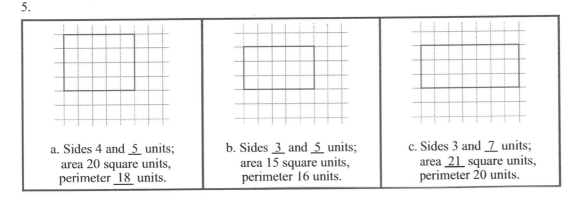

a. Sides 4 and _5_ units; area 20 square units, perimeter _18_ units.	b. Sides _3_ and _5_ units; area 15 square units, perimeter 16 units.	c. Sides 3 and _7_ units; area _21_ square units, perimeter 20 units.

6. 516 mm ÷ 6 = 86 mm

7. First divide the building into two rectangles. That can be done in two different ways.
   One way is shown on the right.
   a. The number sentence is then: 18 m × 33 m + 42 m × 18 m.

   b. The area is 18 m × 33 m + 42 m × 18 m
      = 594 m^2 + 756 m^2 = 1,350 m^2.

   c. The area of the whole plot of land is 80 m × 35 m = 2,800 m^2 .
      We subtract the area of the school to get the area of the actual yard:
      2,800 m^2 − 1,350 m^2 = 1,450 m^2 .

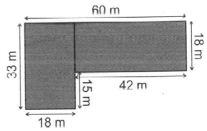

8. a. & b.

	Length	Width	Area	Fencing needed
Sheepyard 1	20 m	20 m	400 m^2	80 m
Sheepyard 2	10 m	40 m	400 m^2	100 m
Sheepyard 3	5 m	80 m	400 m^2	170 m

9. For example, 5 m by 10 m pen will work. Or, 6 m by 9 m. Or, 7 m by 8 m.

Puzzle corner. The width of the inside rectangle is 19 cm − 2 cm − 2 cm = 15 cm. The height of the inner rectangle is 14 cm − 2 cm − 2 cm = 10 cm. The area is then 15 cm × 10 cm = 150 cm^2.

Lines, Rays, and Angles, p. 88

1. a. Ray $\overrightarrow{AB}$ b. angle TAC   c. line segment $\overline{MN}$ d. line $\overleftrightarrow{VW}$ e. angle MAS   f. ray $\overrightarrow{ST}$

2. a. angle EDF or angle FDE   b. angle ACE or angle ECA   c. either a line segment or a ray

3. Answers vary. Check students' answers. For example:

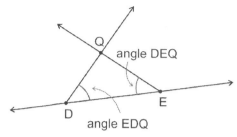

4. a. The second angle is bigger.   b. the second angle   c. the first angle
   d. the second angle   e. the first angle   f. the second angle

5. Check students' work. For example:

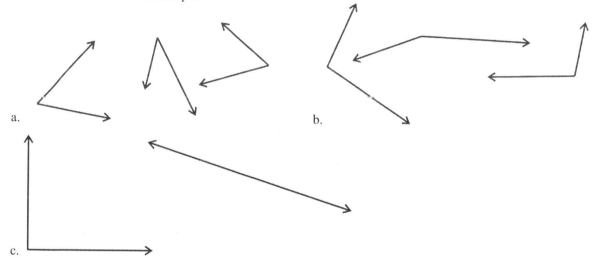

6. a. acute   b. right   c. straight   d. right   e. acute   f. obtuse   g. acute   h. acute   i. obtuse

7. triangle b;   triangle a

Measuring Angles, p. 93

1. a. 35°   b. 72°   c. 18°   d. 50°

2. a. 75°; acute   b. 100°; obtuse   c. 144°; obtuse   d. 135°; obtuse   e. 173°; obtuse   f. 93°; obtuse

3. It is 34°. To get it, subtract 146° from 180°.

4. a. 70°   b. 45°   c. 148°   d. 125°   e. 76°   f. 107°   g. 14°

5. 45°, 102°, 33°. The sum is 180°.

6. You should get 360° if you measure accurately.

Drawing Angles, p. 98

1. Check students' answers.

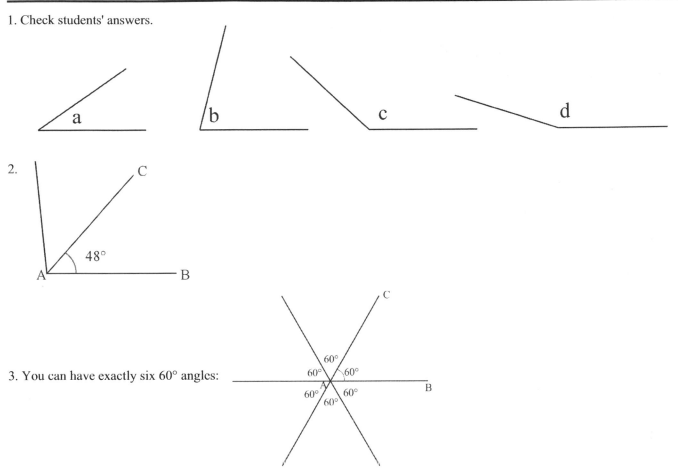

2.

3. You can have exactly six 60° angles:

Angle Problems, p. 100

1. Angle ABC is a __right__ angle, so it measures 90°.

$\angle$ ABD = __25__ ° $\angle$ DBC = __65__ °

2. a. 74°   b. 58°   c. 63°

3. a. 65°   b. 35°   c. 45°

4. a. $38° + x = 72°$; $x = 34°$ b. $37° + x = 90°$; $x = 53°$
   c. $47° + 23° + x = 122°$; $x = 52°$ d. $34° + x = 105°$; $x = 71°$

Measure the two angles. What is the sum of their measures?

$\angle$BAC = __115__ °

$\angle$CAD = __65__ °

Their sum: __180__ °

Angle BAD is called
a __straight__ angle,
and it measures __180__ °.

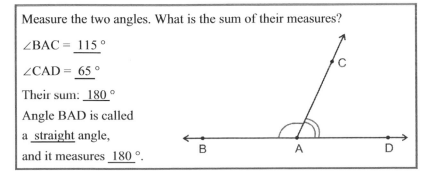

---

5. a. 68°   b. 88°

6. $29° + x = 180°$; $x = 151°$ b. $x + 54° = 180°$; $x = 126°$

73

Angle Problems, continued

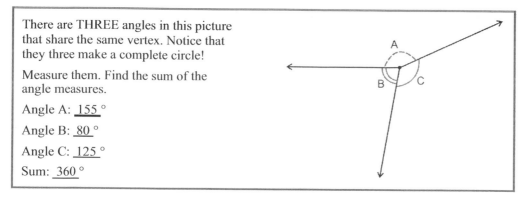

There are THREE angles in this picture that share the same vertex. Notice that they three make a complete circle!

Measure them. Find the sum of the angle measures.

Angle A: __155__ °

Angle B: __80__ °

Angle C: __125__ °

Sum: __360__ °

7. a. ∠A = 115°; ∠B = 85°; ∠C = 130°; ∠D = 30°
   b. 79°; 154°; 127°   c. 129°; 51°; 129°; 51°

8. a. 28° + 116° + 54° + x = 360°;   x = 162°
   b. 57° + 113° + x = 360°;   x = 190°.

9. Measure first the smaller angle that you can measure with the protractor, marked with a dashed line. It is 102°. Then subtract that from 360°: 360° − 102° = 258°. That is the measurement of the angle in question.

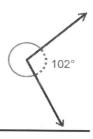

Puzzle corner. a. It is 360° ÷ 8 = 45°.   b. 120°. (First divide 360° by six, then double that.)

Estimating Angles, p. 105

1. Answers vary. The following are the exact measurements.
   a. 30°   b. 45°   c. 100°   d. 60°   e. 160°   f. 80°   g. 40°   h. 10°   i. 15°

2. a. 60°   b. 150°
   c. 1:22; 1:54; 2:27; 3:00, 3:33; 4:05; 4:38; 5:11; 5:44; 6:16; 6:49; 7:21;7:55;
      8:27; 9:00; 9:32; 10:05; 10:38; 11:11; 11:44; 12:16; 12:48

3. Answers vary. The following are the exact measurements. a. 120°   b. 270°   c. 210°

4. Answers vary. The following are the exact measurements.
   a. 110°, 30°, 40°   b. 60°, 30°, 90°   c. 20°, 70°, 90°.

5. Answers vary.

6.

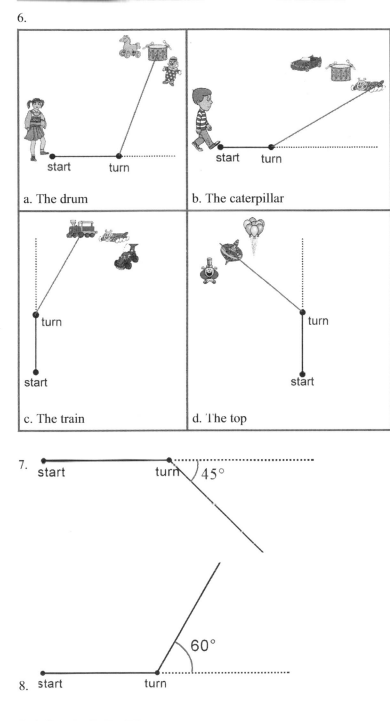

a. The drum

b. The caterpillar

c. The train

d. The top

7. start   turn   45°

8. start   turn   60°

9. A SUPER TURNER

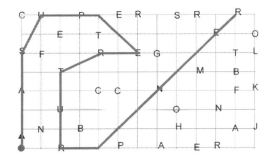

Parallel and Perpendicular Lines, p. 110

1. They are not parallel. They intersect.

2. a. Line segments AB and BC are perpendicular .
      Line segments AD and BC are parallel .

   b. Line segments EF and GH are parallel .
      Line segments EH and FG are parallel .

3.

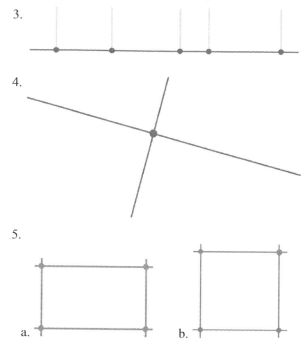

4.

5.

a.   b.

6. Answers vary. Check students' answers.

7. Answers vary. For example:

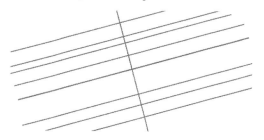

8. Answers vary. Check students' answers. For example:

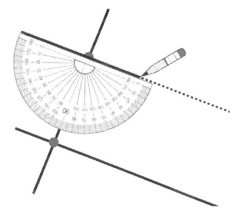

9. The image is not to scale.

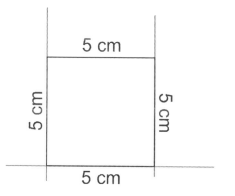

5 cm

5 cm   5 cm

5 cm

10. a. AB ∥ FE, AF ∥ BE, AF ⊥ BF, BF ⊥ BE

    b. $\overline{AB}$ ⊥ s, r ⊥ t, t ⊥ u, r ∥ u

    c. s ⊥ t, s ⊥ $\overline{BC}$, s ⊥ $\overline{FE}$, $\overline{AF}$ ∥ $\overline{CD}$,
       $\overline{AB}$ ∥ $\overline{DE}$, $\overline{BC}$ ∥ $\overline{EF}$, $\overline{EF}$ ∥ t, $\overline{BC}$ ∥ t.

1. a. Answers vary. For example:

   b. Answers vary, but the opposite sides should measure the same.

2. Angle A 68°, Angle B 112°, Angle C 68°, Angle D 112°. You can notice that the opposite corners have the same angle measure. Also, the sum of the angle measures for two neighboring corners is a straight angle (180°).

3. Answers vary. Check students' answers. The opposite sides should measure the same (be congruent), as also the opposite corners.

4. Yes. A rectangle is a parallelogram, because its opposite sides are parallel. A square is also a parallelogram.

5. In the parallelogram ABCD:

   If the side AB is 5 cm, then the side _CD_ is also 5 cm.
   If the side BC is 3 cm, then the side _AD_ is also _3_ cm.
   If the angle at B is 60°, then the angle at _D_ is also 60°.
   If the angle at C is 120°, then the angle at _A_ is also _120_°.

6. a. trapezoid   b. parallelogram   c. trapezoid   d. rhombus

7. A: trapezoid   B: parallelogram   C: trapezoid
   D: rectangle   E: rhombus   F: (no special name except a *scalene quadrilateral*)

8. a. The image is not to scale:

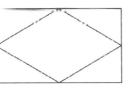

   b. a parallelogram

Triangles, p. 118

1. a. and b. Use a protractor or a triangular ruler to draw the
      right angle.
   c. The other two angles in a right triangle are acute.

> **A right triangle has one right angle.**
> **The other two angles are _acute_.**

2. c. In an obtuse triangle the other two angles are acute.

> **An obtuse triangle has one obtuse angle.**
> **The other two angles are _acute_.**

3. a. b. Answers vary. Check students' answers.

4.
> **Right triangles** have exactly 1 _right angle_,
> and the other two angles are _acute_.
>
> **Obtuse triangles** have exactly 1 _obtuse angle_,
> and the other two angles are _acute_.
>
> **Acute triangles** have _3_ _acute_ angles.

5. a. obtuse   b. acute   c. acute
   d. obtuse   e. acute   f. right
   g. triangle ABD is right, triangle BCD is right,
      and triangle ACD is obtuse

6. a. acute   b. right   c. obtuse   d. acute
   e. the black triangle is obtuse. The red triangle is acute.

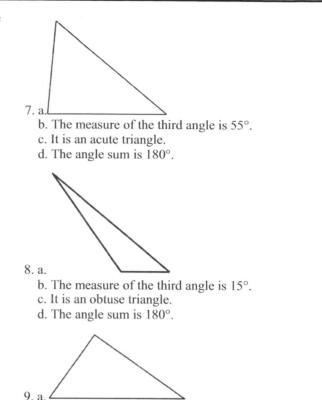

7. a.
   b. The measure of the third angle is 55°.
   c. It is an acute triangle.
   d. The angle sum is 180°.

8. a.
   b. The measure of the third angle is 15°.
   c. It is an obtuse triangle.
   d. The angle sum is 180°.

9. a.
   b. The measure of the third angle is 90°.
   c. It is a right triangle.
   d. The angle sum is 180°.

Line Symmetry, p. 122

1. a. no   b. yes   c. yes   d. no   e. no   f. no   g. no   h. yes   i. yes

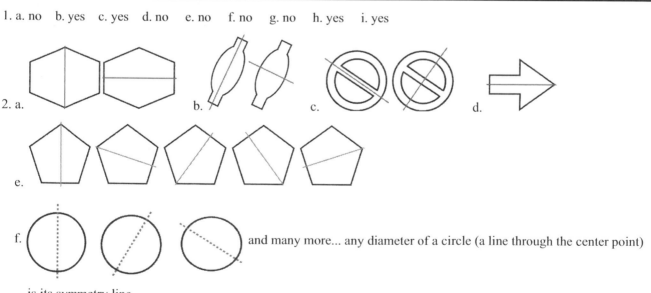

2. a.   b.   c.   d.

e.

f.   and many more... any diameter of a circle (a line through the center point)

   is its symmetry line.

3. You can draw a vertical symmetry line to the letters A, H, I, M, O, T, U, V, W, X, and Y.
   You can draw a horizontal symmetry line to the letters B, C, D, E, H, I, K, O, and X.

Line Symmetry, continued

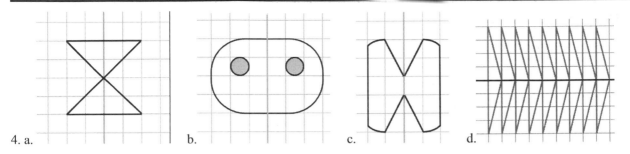

4. a.   b.   c.   d.

Mixed Review, p. 125

1. a. 3,000 g;   7,400 g   b. 5,000 ml;   2,060 ml   c. 9,000 m;   4,250 m

2. 1 1/2 liters are left now. Six glasses got filled.

3. a. 48 in.;   74 in.   b. 24 fl. oz.;   8 qt   c. 64 oz;   121 oz

4. 1 pint bottle is more. It is 4 ounces more.

5. a. 8 in.   b. 24 in. or 2 ft   c. 8 ft

6. a. 960   b. 508   c. 1,670   d. 1,099

7. a.

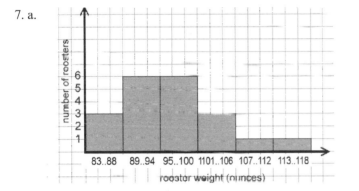

Weight (ounces)	Frequency
83..88	3
89..94	6
95..100	6
101..106	3
107..112	1
113..118	1

  b. 95 1/2 ounces. You can see this in the bar graph because the number 95 1/2 is near the middle and near the peak of the graph. You could also see it from the data itself, noting that lots of the chicken weights are 90-something.

8. a. The silverware set costs $4 × $13 = $52. The two items together cost $13 + $52 = $65.
   b. He still has $200 − 8 × $18 = $56.
   c. From 22:15 till 7:00 is 8 hours 45 minutes. But he did not sleep from 3:30 till 5:10, which is 1 hour 40 minutes. So, we subtract and get that he slept 8 h 45 min − 1 h 40 min = 7 h 5 min.

1. Side 1: 17 1/2 ft; Side 2: 10 ft; Side 3: 17 1/2 ft

2. The area is 320 m².
   First, divide the shape into two rectangles. You also need to use subtraction to find some side lengths.

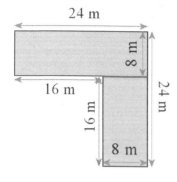

The upper rectangle is 24 m × 8 m so its area is 192 m².
The lower rectangle is 8 m × 16 m so its area is 128 m².
In total, the area is 320 m².

3. a. right   b. acute   c. obtuse   d. acute   e. acute
   f. right   g. right   h. obtuse   i. right   j. obtuse
   k. obtuse

4. Check students' answers. Here is one such angle:

5. It is 84°. The student can figure it out using any method, but if he/she writes and equation, it is
   $64° + x + 32° = 180°$; $x = 84°$.

6.

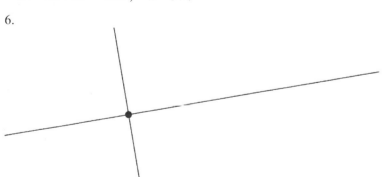

7. b. is possible to do. Answers will vary as the other two angles can vary. For example:

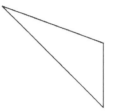

8. Check students' answers.

9. Check students' answers. The angle sum should be 180° or close to that.

10. right triangles.

11. AB∥n, CD ∥ m, AC ∥ BD

    AC ⊥ m, BD ⊥ m

12. a. b. answers vary. In b. the angle measures of the opposite corners should be the same, and each two neighboring corners should have an angle sum of 180° (or close).

13. a.   b. not symmetrical   c.

    d. not symmetrical   e.

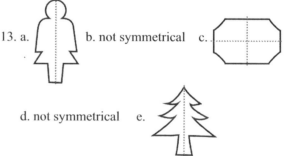

Chapter 7: Fractions

One Whole and its Fractional Parts, p. 135

1.

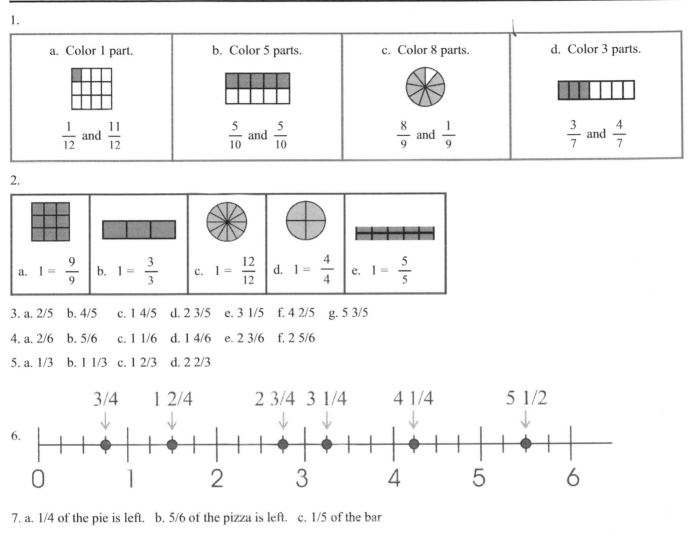

a. Color 1 part.	b. Color 5 parts.	c. Color 8 parts.	d. Color 3 parts.
$\frac{1}{12}$ and $\frac{11}{12}$	$\frac{5}{10}$ and $\frac{5}{10}$	$\frac{8}{9}$ and $\frac{1}{9}$	$\frac{3}{7}$ and $\frac{4}{7}$

2.

a. $1 = \frac{9}{9}$	b. $1 = \frac{3}{3}$	c. $1 = \frac{12}{12}$	d. $1 = \frac{4}{4}$	e. $1 = \frac{5}{5}$

3. a. 2/5   b. 4/5   c. 1 4/5   d. 2 3/5   e. 3 1/5   f. 4 2/5   g. 5 3/5

4. a. 2/6   b. 5/6   c. 1 1/6   d. 1 4/6   e. 2 3/6   f. 2 5/6

5. a. 1/3   b. 1 1/3   c. 1 2/3   d. 2 2/3

6.

7. a. 1/4 of the pie is left.   b. 5/6 of the pizza is left.   c. 1/5 of the bar

8.

a. Color 1 part.	b. Color 10 parts.	c. Color 3 parts.	d. Color 15 parts.
$\frac{1}{6} + \frac{5}{6} = 1$	$\frac{10}{12} + \frac{2}{12} = 1$	$\frac{3}{8} + \frac{5}{8} = 1$	$\frac{15}{100} + \frac{85}{100} = 1$

9. a. 1/4   b. 1/7   c. 7/8   d. 1/12

10. a. 2/4 or 1/2 liter   b. 14/20 of the bread is left.

11. a. 1/10 of 90 km = 9 km. Then, 4/10 of 90 km = 36 km.
    b. First, divide $45.50 into five parts: $45.50 ÷ 5 = $9.10. Cindy pays 2/5 of the bill, or double that, which is $18.20. Sandy pays the rest, or $27.30.
    c. 7/9 is left. $2,100 is left. One-ninth of his paycheck is $300, so seven-ninths of it is 7 × $300 = $2,100.

1. a. 2 3/4   b. 1 1/2   c. 4 2/10   d. 8 1/3   e. 2 4/9   f. 3 5/6

2.

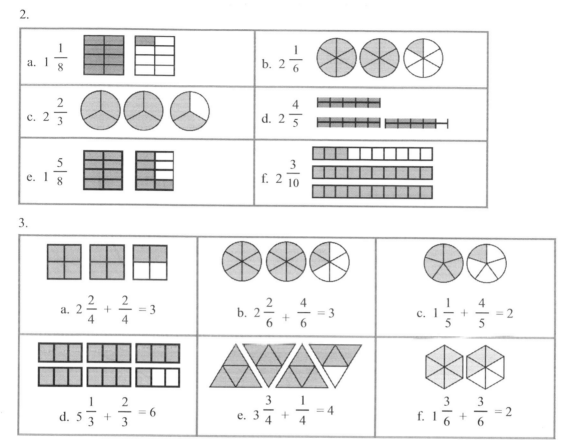

a. $1\frac{1}{8}$	b. $2\frac{1}{6}$
c. $2\frac{2}{3}$	d. $2\frac{4}{5}$
e. $1\frac{5}{8}$	f. $2\frac{3}{10}$

3.

a. $2\frac{2}{4} + \frac{2}{4} = 3$

b. $2\frac{2}{6} + \frac{4}{6} = 3$

c. $1\frac{1}{5} + \frac{4}{5} = 2$

d. $5\frac{1}{3} + \frac{2}{3} = 6$

e. $3\frac{3}{4} + \frac{1}{4} = 4$

f. $1\frac{3}{6} + \frac{3}{6} = 2$

4. a. 3/4   b. 8/10   c. 5/9   d. 7/8

5. Three-fourths of a cup will finish filling the pitcher.

6. Your train of cars would be 4 1/2 inches long.

7. She needs five scoops of flour.

8. There is two-thirds of a pound of extra beef.

9. He had 1 7/12 of the bread left.

10.

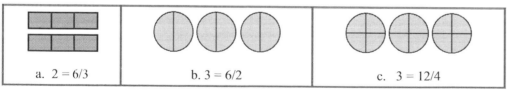

a. 2 = 6/3	b. 3 = 6/2	c. 3 = 12/4

11.

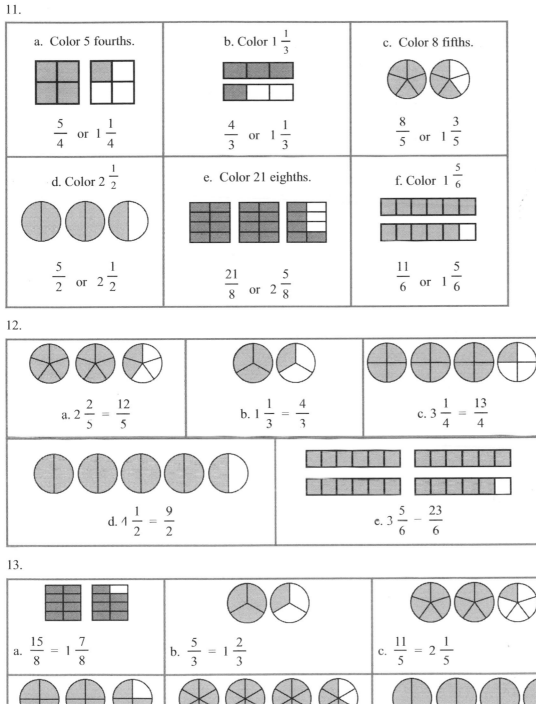

a. Color 5 fourths.

$$\frac{5}{4} \quad \text{or} \quad 1\frac{1}{4}$$

b. Color $1\frac{1}{3}$

$$\frac{4}{3} \quad \text{or} \quad 1\frac{1}{3}$$

c. Color 8 fifths.

$$\frac{8}{5} \quad \text{or} \quad 1\frac{3}{5}$$

d. Color $2\frac{1}{2}$

$$\frac{5}{2} \quad \text{or} \quad 2\frac{1}{2}$$

e. Color 21 eighths.

$$\frac{21}{8} \quad \text{or} \quad 2\frac{5}{8}$$

f. Color $1\frac{5}{6}$

$$\frac{11}{6} \quad \text{or} \quad 1\frac{5}{6}$$

12.

a. $2\frac{2}{5} = \frac{12}{5}$

b. $1\frac{1}{3} = \frac{4}{3}$

c. $3\frac{1}{4} = \frac{13}{4}$

d. $4\frac{1}{2} = \frac{9}{2}$

e. $3\frac{5}{6} = \frac{23}{6}$

13.

a. $\frac{15}{8} = 1\frac{7}{8}$

b. $\frac{5}{3} = 1\frac{2}{3}$

c. $\frac{11}{5} = 2\frac{1}{5}$

d. $\frac{11}{4} = 2\frac{3}{4}$

e. $\frac{22}{6} = 3\frac{4}{6}$

f. $\frac{7}{2} = 3\frac{1}{2}$

14.

a. $\frac{21}{4}$	b. $\frac{19}{3}$	c. $\frac{26}{3}$	d. $\frac{22}{5}$
e. $\frac{81}{10}$	f. $\frac{47}{9}$	g. $\frac{37}{10}$	h. $\frac{32}{9}$

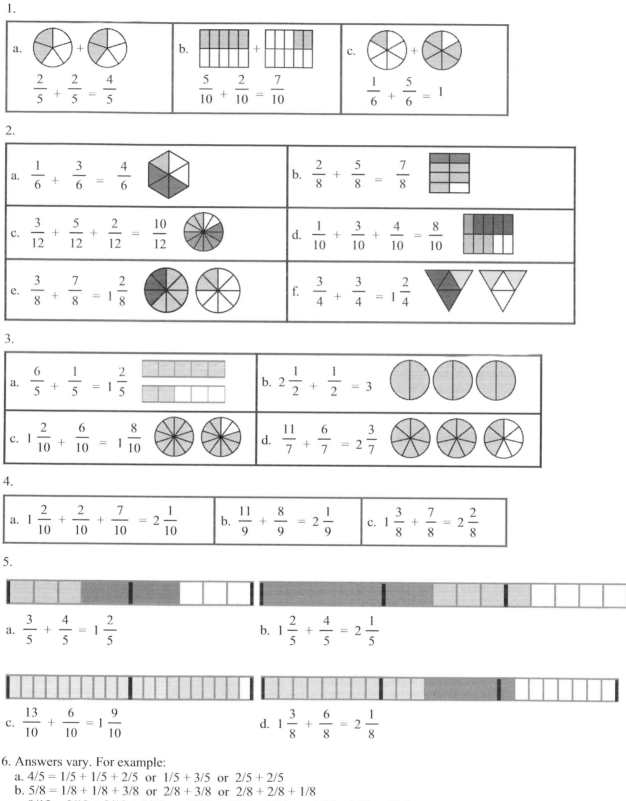

1.

a. $\dfrac{2}{5} + \dfrac{2}{5} = \dfrac{4}{5}$

b. $\dfrac{5}{10} + \dfrac{2}{10} = \dfrac{7}{10}$

c. $\dfrac{1}{6} + \dfrac{5}{6} = 1$

2.

a. $\dfrac{1}{6} + \dfrac{3}{6} = \dfrac{4}{6}$

b. $\dfrac{2}{8} + \dfrac{5}{8} = \dfrac{7}{8}$

c. $\dfrac{3}{12} + \dfrac{5}{12} + \dfrac{2}{12} = \dfrac{10}{12}$

d. $\dfrac{1}{10} + \dfrac{3}{10} + \dfrac{4}{10} = \dfrac{8}{10}$

e. $\dfrac{3}{8} + \dfrac{7}{8} = 1\dfrac{2}{8}$

f. $\dfrac{3}{4} + \dfrac{3}{4} = 1\dfrac{2}{4}$

3.

a. $\dfrac{6}{5} + \dfrac{1}{5} = 1\dfrac{2}{5}$

b. $2\dfrac{1}{2} + \dfrac{1}{2} = 3$

c. $1\dfrac{2}{10} + \dfrac{6}{10} = 1\dfrac{8}{10}$

d. $\dfrac{11}{7} + \dfrac{6}{7} = 2\dfrac{3}{7}$

4.

a. $1\dfrac{2}{10} + \dfrac{2}{10} + \dfrac{7}{10} = 2\dfrac{1}{10}$

b. $\dfrac{11}{9} + \dfrac{8}{9} = 2\dfrac{1}{9}$

c. $1\dfrac{3}{8} + \dfrac{7}{8} = 2\dfrac{2}{8}$

5.

a. $\dfrac{3}{5} + \dfrac{4}{5} = 1\dfrac{2}{5}$

b. $1\dfrac{2}{5} + \dfrac{4}{5} = 2\dfrac{1}{5}$

c. $\dfrac{13}{10} + \dfrac{6}{10} = 1\dfrac{9}{10}$

d. $1\dfrac{3}{8} + \dfrac{6}{8} = 2\dfrac{1}{8}$

6. Answers vary. For example:
   a. 4/5 = 1/5 + 1/5 + 2/5 or 1/5 + 3/5 or 2/5 + 2/5
   b. 5/8 = 1/8 + 1/8 + 3/8 or 2/8 + 3/8 or 2/8 + 2/8 + 1/8
   c. 9/12 = 3/12 + 3/12 + 3/12 or 2/12 + 5/12 + 2/12 or 1/12+ 4/12 + 4/12
   d. 4/3 = 2/3 + 2/3 or 1/3 + 3/3
   e. 9/6 = 1/6 + 8/6 or 2/6 + 7/6 or 3/6 + 6/6

7.

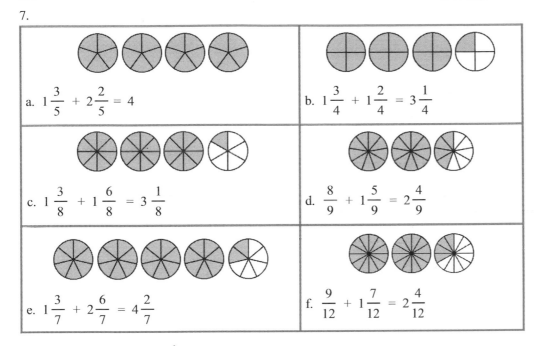

a. $1\dfrac{3}{5} + 2\dfrac{2}{5} = 4$

b. $1\dfrac{3}{4} + 1\dfrac{2}{4} = 3\dfrac{1}{4}$

c. $1\dfrac{3}{8} + 1\dfrac{6}{8} = 3\dfrac{1}{8}$

d. $\dfrac{8}{9} + 1\dfrac{5}{9} = 2\dfrac{4}{9}$

e. $1\dfrac{3}{7} + 2\dfrac{6}{7} = 4\dfrac{2}{7}$

f. $\dfrac{9}{12} + 1\dfrac{7}{12} = 2\dfrac{4}{12}$

8. Answers vary. For example:
   a. 1 3/10 = 1 + 3/10 or 5/10 + 8/10 or 4/10 + 9/10 or 2/10 + 11/10

   b. 3 1/5 = 4/5 + 12/5 or 10/5 + 6/5 or 8/5 + 8/5 or 1/5 + 15/5

Adding Fractions and Mixed Numbers 2, p. 146

1. a. 2/6   b. 1   c. 7/8   d. 1 2/5   e. 2   f. 1 8/10   g. 3

2. a. 3 1/5   b. 3 1/5   c. 3 2/10   d. 3 7/10

3. a. 5   b. 7 1/6   c. 13 1/4   d. 10 2/8

4. a. 3/5   b. 9/12   c. 3/6   d. 1 7/8   e. 1 3/8   f. 2 2/4

5. These answers can be fixed in different ways. For example:

a. In the first one, Emma has one seventh too much. The second one is correct.	b. In the first one, Peter is lacking one third from the addition. In the second one, he has one third too much.
$1\dfrac{5}{7} = \dfrac{2}{7} + 1\dfrac{1}{7} + \dfrac{2}{7}$	$2\dfrac{1}{3} = \dfrac{2}{3} + \dfrac{2}{3} + \dfrac{2}{3} + \dfrac{1}{3}$
$1\dfrac{5}{7} = \dfrac{10}{7} + \dfrac{2}{7}$	$2\dfrac{1}{3} = \dfrac{5}{3} + \dfrac{2}{3}$

6. a. 1 1/2 + 1/2 + 1/2 = 2 1/2. The recipe calls for 2 1/2 cups of flour.
   b. 3/12 + 2/12 + 4/12 = 9/12. The children ate 9/12 of the chocolate bar. There is 3/12 left.
   c. 1 3/4 + 1 1/4 = 3. They took three hours.
   d. 1 1/2 + 3/4 = 2. He drank 2 cups of liquid.

7. 2 1/4 + 3 1/4 + 2 1/4 + 3 1/4 = 11. Its perimeter is 11 inches.

8. 2 3/8 + 2 3/8 + 2 3/8 = 6 9/8 = 7 1/8. Its perimeter is 7 1/8 inches.

Adding Fractions and Mixed Numbers 2, continued

9.

> A birthday cake
> 8 eggs
> 1 1/2 cups sugar
> 2 1/2 cups flour
> 3 tsp baking powder
> 2 cups whipped cream
> sliced fruit

Puzzle corner:
a. 1 1/5   b. 1 3/4   c. 1 3/6

Equivalent Fractions, p. 149

1.

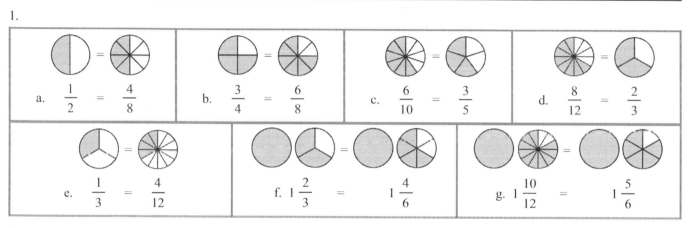

a. $\dfrac{1}{2} = \dfrac{4}{8}$ b. $\dfrac{3}{4} = \dfrac{6}{8}$ c. $\dfrac{6}{10} = \dfrac{3}{5}$ d. $\dfrac{8}{12} = \dfrac{2}{3}$

e. $\dfrac{1}{3} = \dfrac{4}{12}$ f. $1\dfrac{2}{3} = 1\dfrac{4}{6}$ g. $1\dfrac{10}{12} = 1\dfrac{5}{6}$

2.

a. $\dfrac{3}{3} = \dfrac{6}{6}$	b. $\dfrac{4}{3} = \dfrac{8}{6}$	c. $\dfrac{7}{3} = \dfrac{14}{6}$
d. $2\dfrac{1}{3} = 2\dfrac{2}{6}$	e. $1\dfrac{2}{3} = 1\dfrac{4}{6}$	f. $2\dfrac{2}{3} = 2\dfrac{4}{6}$

3.

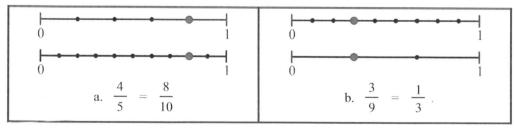

a. $\dfrac{4}{5} = \dfrac{8}{10}$ b. $\dfrac{3}{9} = \dfrac{1}{3}$

4.

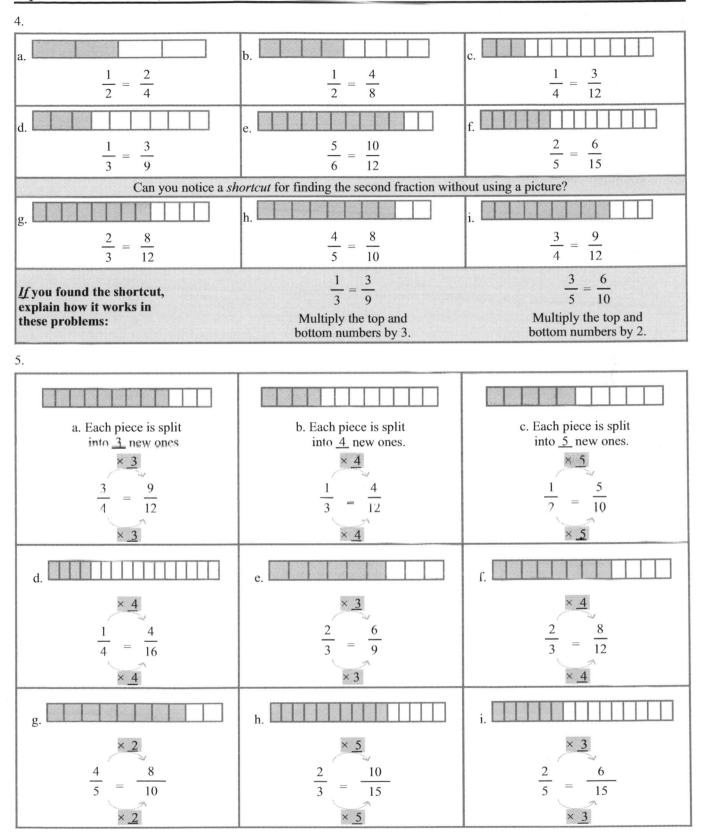

a. $\frac{1}{2} = \frac{2}{4}$	b. $\frac{1}{2} = \frac{4}{8}$	c. $\frac{1}{4} = \frac{3}{12}$
d. $\frac{1}{3} = \frac{3}{9}$	e. $\frac{5}{6} = \frac{10}{12}$	f. $\frac{2}{5} = \frac{6}{15}$

Can you notice a *shortcut* for finding the second fraction without using a picture?

g. $\frac{2}{3} = \frac{8}{12}$ h. $\frac{4}{5} = \frac{8}{10}$ i. $\frac{3}{4} = \frac{9}{12}$

_If_ you found the shortcut, explain how it works in these problems:

$\frac{1}{3} = \frac{3}{9}$
Multiply the top and bottom numbers by 3.

$\frac{3}{5} = \frac{6}{10}$
Multiply the top and bottom numbers by 2.

5.

a. Each piece is split into __3__ new ones
$\times \underline{3}$
$\frac{3}{4} = \frac{9}{12}$
$\times \underline{3}$

b. Each piece is split into __4__ new ones.
$\times \underline{4}$
$\frac{1}{3} = \frac{4}{12}$
$\times \underline{4}$

c. Each piece is split into __5__ new ones.
$\times \underline{5}$
$\frac{1}{2} = \frac{5}{10}$
$\times \underline{5}$

d. $\times \underline{4}$
$\frac{1}{4} = \frac{4}{16}$
$\times \underline{4}$

e. $\times \underline{3}$
$\frac{2}{3} = \frac{6}{9}$
$\times \underline{3}$

f. $\times \underline{4}$
$\frac{2}{3} = \frac{8}{12}$
$\times \underline{4}$

g. $\times \underline{2}$
$\frac{4}{5} = \frac{8}{10}$
$\times \underline{2}$

h. $\times \underline{5}$
$\frac{2}{3} = \frac{10}{15}$
$\times \underline{5}$

i. $\times \underline{3}$
$\frac{2}{5} = \frac{6}{15}$
$\times \underline{3}$

6. a. 15/18   b. 15/20   c. 8/20   d. 90/100

7.

a. Pieces were split into 3 new ones. $\frac{1}{2} = \frac{3}{6}$	b. Pieces were split into 10 new ones. $\frac{3}{10} = \frac{30}{100}$	c. Pieces were split into 6 new ones. $\frac{2}{5} = \frac{12}{30}$	d. Pieces were split into 5 new ones. $\frac{7}{8} = \frac{35}{40}$
e. $\frac{2}{3} = \frac{4}{6}$	f. $\frac{3}{5} = \frac{9}{15}$	g. $\frac{5}{6} = \frac{10}{12}$	h. $\frac{1}{3} = \frac{3}{9}$

8.

a. $\frac{1}{10} = \frac{10}{100}$	b. $\frac{3}{10} = \frac{30}{100}$	c. $\frac{6}{10} = \frac{60}{100}$	d. $\frac{4}{10} = \frac{40}{100}$	e. $\frac{13}{10} = \frac{130}{100}$

9.

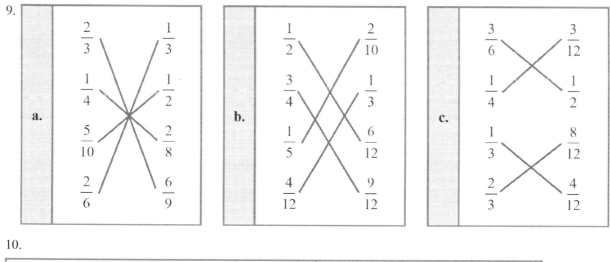

10.

a. $\frac{1}{2} = \frac{2}{4} = \frac{3}{6} = \frac{4}{8} = \frac{5}{10} = \frac{6}{12} = \frac{7}{14}$	b. $\frac{1}{3} = \frac{2}{6} = \frac{3}{9} = \frac{4}{12} = \frac{5}{15}$

11. a. 18/100   b. 73/100   c. 75/100   d. 99/100   e. 93/100   f. 114/100 or 1 14/100
    g. 147/100 or 1 47/100   h. 3 78/100   i. 102/100 or 1 2/100

12. Answers vary. For example:

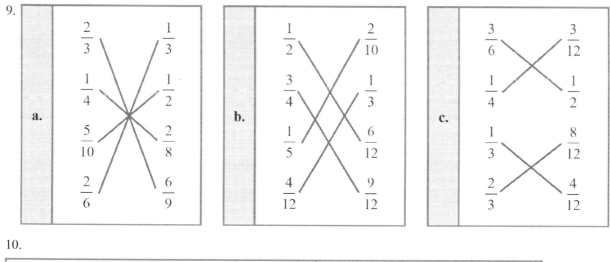

$$\frac{1}{3} = \frac{4}{12}$$

Puzzle corner:

a. $\frac{3}{4} + \frac{1}{2}$ $\downarrow$ $\downarrow$ $\frac{3}{4} + \frac{2}{4} = \frac{5}{4} = 1\frac{1}{4}$	b. $\frac{1}{5} + \frac{3}{10}$ $\downarrow$ $\downarrow$ $\frac{2}{10} + \frac{3}{10} = \frac{5}{10}$	c. $\frac{2}{3} + \frac{2}{9}$ $\downarrow$ $\downarrow$ $\frac{6}{9} + \frac{2}{9} = \frac{8}{9}$

1. a. 8/10   b. 2 2/6   c. 1 2/9   d. 2 4/8

2. a. 4/12   b. 2/10   c. 6/4 = 1 2/4   d. 1/8   e. 2 4/8   f. 3 4/12   g. 1 3/6

3. There are 2 9/12 of the pies left.

4. a. 2 1/6   b. 1 6/10   c. 3/8   d. 1 3/5

5. a. 1 1/5   b. 1 2/5   c. 2/5

6.

a. $1\dfrac{2}{10} - \dfrac{6}{10}$	b. $2\dfrac{1}{3} - \dfrac{2}{3}$	c. $1\dfrac{3}{10} - \dfrac{7}{10}$	d. $2\dfrac{2}{5} - \dfrac{4}{5}$
↓ ↓	↓ ↓	↓ ↓	↓ ↓
$\dfrac{12}{10} - \dfrac{6}{10} = \dfrac{6}{10}$	$\dfrac{7}{3} - \dfrac{2}{3} = \dfrac{5}{3} = 1\dfrac{2}{3}$	$\dfrac{13}{10} - \dfrac{7}{10} = \dfrac{6}{10}$	$\dfrac{12}{5} - \dfrac{4}{5} = \dfrac{8}{5} = 1\dfrac{3}{5}$

7. a. 1 1/4   b. 5 1/9   c. 3 1/2   d. 7 2/5

8.

a. $\dfrac{6}{10} - \dfrac{15}{100}$	b. $\dfrac{7}{10} - \dfrac{38}{100}$	c. $\dfrac{54}{100} - \dfrac{2}{10}$
↓ ↓	↓ ↓	↓ ↓
$\dfrac{60}{100} - \dfrac{15}{100} = \dfrac{45}{100}$	$\dfrac{70}{100} - \dfrac{38}{100} = \dfrac{32}{100}$	$\dfrac{54}{100} - \dfrac{20}{100} = \dfrac{34}{100}$

9. It is 1 inch. Half of the perimeter is 2 1/2 inches, and the two sides add up to that (1 in. + 1 1/2 in. = 2 1/2 in.).

10. 6 − 2 2/3 = 3 1/3. The part left is 3 1/3 yards long. In feet, 2 2/3 yards = 8 feet, and 3 1/3 feet = 10 feet.

11. #1: You can change 3 1/5 and 1 4/5 into fractions and get 16/5 and 9/5. Subtracting those, we get 7/5, which is 1 2/5.
    #2: Subtract in parts: First subtract 3 1/5 − 1, which is 2 1/5. Then subtract 2 1/5 − 4/5 = 1 2/5.

12. a. 1 2/4   b. 2 6/8   c. 3 2/6   d. 4/5   e. 4 3/5   f. 2 2/3

13. 11/12 of a pizza is left. Edward, Abigail, Jack, and John ate 13 pieces, which is 1 1/12 of a pizza. Since Mom and Dad ate 1 pizza, in total 2 1/12 of the pizzas were consumed. So, 11/12 of a pizza is left.

14. a. 6 − 2 2/3 = 3 1/3. There is 3 1/3 cups of flour left.   b. She can make one more batch.

Puzzler corner. 1 3/4 + (?) = 3 1/4. (?) = 1 2/4 or 1 1/2. The other side is 1 1/2 inches.

1. a. $<$ b. $>$ c. $<$ d. $>$

2. a. $>$ b. $>$ c. $>$ d. $<$

3. a. $<$ b. $>$ c. $>$ d. $>$ e. $>$ f. $<$ g. $=$ h. $<$

4. a. $\dfrac{3}{8}, \dfrac{3}{6}, \dfrac{6}{8}$ b. $\dfrac{2}{5}, \dfrac{5}{6}, \dfrac{6}{5}$ c. $\dfrac{1}{7}, \dfrac{1}{4}, \dfrac{5}{8}$

5. a. $>$ b. $<$ c. $<$ d. $<$

6. a. $<$ b. $<$ c. $<$ d. $<$ e. $>$ f. $<$ g. $>$ h. $>$ i. $>$ j. $>$

7.

a. $\dfrac{1}{5} \quad \dfrac{3}{10}$ $\downarrow \quad \downarrow$ $\dfrac{2}{10} < \dfrac{3}{10}$	b. $\dfrac{3}{4} \quad \dfrac{5}{8}$ $\downarrow \quad \downarrow$ $\dfrac{6}{8} > \dfrac{5}{8}$	c. $\dfrac{5}{12} \quad \dfrac{1}{3}$ $\downarrow \quad \downarrow$ $\dfrac{5}{12} > \dfrac{4}{12}$	d. $\dfrac{11}{12} \quad \dfrac{5}{6}$ $\downarrow \quad \downarrow$ $\dfrac{11}{12} > \dfrac{10}{12}$
e. $\dfrac{3}{4} \quad \dfrac{9}{12}$ $\downarrow \quad \downarrow$ $\dfrac{9}{12} = \dfrac{9}{12}$	f. $\dfrac{5}{9} \quad \dfrac{2}{3}$ $\downarrow \quad \downarrow$ $\dfrac{5}{9} < \dfrac{6}{9}$	g. $\dfrac{1}{3} \quad \dfrac{2}{9}$ $\downarrow \quad \downarrow$ $\dfrac{3}{9} > \dfrac{2}{9}$	h. $\dfrac{3}{12} \quad \dfrac{1}{3}$ $\downarrow \quad \downarrow$ $\dfrac{3}{12} < \dfrac{4}{12}$

8. a. cannot compare   b. $3/9 = 2/6$ c. $7/10 > 5/8$ d. cannot compare   e. cannot compare   f. cannot compare

9. $\dfrac{1}{3}, \dfrac{3}{8}, \dfrac{2}{5}, \dfrac{5}{8}, \dfrac{2}{3}$

10. a. Answers vary. For example: $<$

   b. Answers vary. For example: $>$

11. Angie ate more pizza. She ate 1/8 of the pizza more than Joe. That is because Joe ate $1/4 = 2/8$ of the pizza.

12. Chloe does. She pays 3/10, which is 30/100, of her paycheck in taxes.

13. If it is discounted 4/10 of its price, because $4/10 = 40/100$.

14. a. The wholes are not the same size.

   b. $>$

15.

a. $\dfrac{3}{7}, \dfrac{3}{5}, 1\dfrac{1}{7}$	b. $\dfrac{3}{8}, \dfrac{3}{6}, 1\dfrac{1}{4}$	c. $\dfrac{4}{9}, \dfrac{2}{3}, \dfrac{6}{5}$

Puzzle Corner. Dad ate more. Eating 2/3 of the smaller pizza, which is 1/2 the size of the larger pizza, is equal to eating 1/3 of the larger pizza. Dad ate 3/8 of the larger pizza. Now, $3/8 > 1/3$ (see exercise #9), so Dad ate more pizza.

1.

a. $\dfrac{3}{7} = 3 \times \dfrac{1}{7}$	b. $\dfrac{6}{9} = 6 \times \dfrac{1}{9}$	c. $4 \times \dfrac{1}{5} = \dfrac{4}{5}$	d. $7 \times \dfrac{1}{10} = \dfrac{7}{10}$

2.

a. $\dfrac{8}{7} = 8 \times \dfrac{1}{7}$	b. $1\dfrac{3}{5} = \dfrac{8}{5} = 8 \times \dfrac{1}{5}$	c. $1\dfrac{2}{3} = \dfrac{5}{3} = 5 \times \dfrac{1}{3}$
d. $10 \times \dfrac{1}{6} = \dfrac{10}{6} = 1\dfrac{4}{6}$	e. $7 \times \dfrac{1}{4} = \dfrac{7}{4} = 1\dfrac{3}{4}$	f. $9 \times \dfrac{1}{3} = \dfrac{9}{3} = 3$

3. a. $10 \times 1/3 = 10/3 = 3\ 1/3$. She needs to buy at least 3 1/3 lb of chicken.
   b. Between 3 and 4.
   c. $10 \times 1/2 = 5$. She needs 5 quarts of juice.

4.

a. $3 \times \dfrac{2}{4} = \dfrac{6}{4} = 1\dfrac{2}{4}$	b. $4 \times \dfrac{2}{6} = \dfrac{8}{6} = 1\dfrac{2}{6}$	c. $2 \times \dfrac{7}{8} = \dfrac{14}{8} = 1\dfrac{6}{8}$

5.

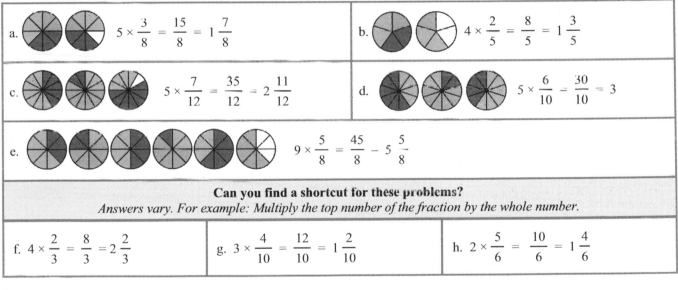

a. $5 \times \dfrac{3}{8} = \dfrac{15}{8} = 1\dfrac{7}{8}$	b. $4 \times \dfrac{2}{5} = \dfrac{8}{5} = 1\dfrac{3}{5}$
c. $5 \times \dfrac{7}{12} = \dfrac{35}{12} = 2\dfrac{11}{12}$	d. $5 \times \dfrac{6}{10} = \dfrac{30}{10} = 3$
e. $9 \times \dfrac{5}{8} = \dfrac{45}{8} - 5\dfrac{5}{8}$	

Can you find a shortcut for these problems?
Answers vary. For example: Multiply the top number of the fraction by the whole number.

f. $4 \times \dfrac{2}{3} = \dfrac{8}{3} = 2\dfrac{2}{3}$	g. $3 \times \dfrac{4}{10} = \dfrac{12}{10} = 1\dfrac{2}{10}$	h. $2 \times \dfrac{5}{6} = \dfrac{10}{6} = 1\dfrac{4}{6}$

6.

a. $\dfrac{8}{5} = 4 \times \dfrac{2}{5}$	b. $\dfrac{9}{4} = 3 \times \dfrac{3}{4}$	c. $2\dfrac{2}{3} = 2 \times 1\dfrac{1}{3}$

7. a. 1 1/4   b. 2   c. 1 1/7   d. 1 4/10   e. 2 2/8   f. 14/100
   g. 2 1/10   h. 72/100   i. 3 3/10   j. 3 1/8   k. 2 2/3   l. 3 2/4

8. $4 \times 7/8$ in. $= 28/8$ in. $= 3\ 4/8$ in. (which is also equal to 3 1/2 in.)

9. a. $5 \times 1\ 1/8$ in. $= 5\ 5/8$ in.
   b. Double the previous result to get 10 10/8 in. = 11 2/8 in.

10. Meat: $8 \times 1/4$ lb $= 2$ lb. Pasta: $8 \times 3/4$ C $= 24/4$ C $= 6$ C.

1.
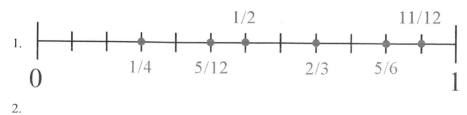

2.

a. $\frac{2}{6}, \frac{1}{2}, \frac{2}{3}$	b. $\frac{1}{8}, \frac{1}{4}, \frac{3}{8}$
c. $\frac{2}{5}, \frac{1}{2}, \frac{3}{5}$	d. $\frac{3}{8}, \frac{3}{4}, \frac{4}{5}$

3. $3 \times 3/5$ mi. $= 9/5$ mi. $= 1\ 4/5$ mi.

4. Answers may vary.

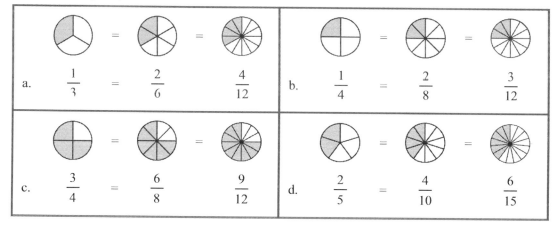

a. $\frac{1}{3} = \frac{2}{6}\qquad \frac{4}{12}$	b. $\frac{1}{4} = \frac{2}{8}\qquad \frac{3}{12}$
c. $\frac{3}{4} = \frac{6}{8}\qquad \frac{9}{12}$	d. $\frac{2}{5} = \frac{4}{10}\qquad \frac{6}{15}$

5. a. 4/7   b. 3 1/3   c. 2 1/4   d. 2   e. 7   f. 2

6. a. 1   b. 1 3/8   c. 7 1/5   d. 8/12   e. 8/10   f. 4 2/4

7. Multiplying by one-half is actually the same as dividing by 2. Fill in.

a.	b.	c.
$2 \times \frac{1}{2} = 1$	$7 \times \frac{1}{2} = 3\frac{1}{2}$	$15 \times \frac{1}{2} = 7\frac{1}{2}$
$3 \times \frac{1}{2} = 1\frac{1}{2}$	$8 \times \frac{1}{2} = 4$	$20 \times \frac{1}{2} = 10$
$4 \times \frac{1}{2} = 2$	$9 \times \frac{1}{2} = 4\frac{1}{2}$	$17 \times \frac{1}{2} = 8\frac{1}{2}$
$5 \times \frac{1}{2} = 2\frac{1}{2}$	$10 \times \frac{1}{2} = 5$	$21 \times \frac{1}{2} = 10\frac{1}{2}$
$6 \times \frac{1}{2} = 3$	$11 \times \frac{1}{2} = 5\frac{1}{2}$	$32 \times \frac{1}{2} = 16$

Puzzle Corner.
a. $1/2 + 3/8 = 4/8 + 3/8 = 7/8$ b. $1/3 + 1/6 = 2/6 + 1/6 = 3/6$ c. $1/3 + 2/9 = 3/9 + 2/9 = 5/9$

Mixed Review, p. 166

1.

a. $57 \div 5 = 11$ R2	b. $34 \div 7 = 4$ R6	c. $33 \div 9 = 3$ R 6
$11 \times 5 + 2 = 57$	$4 \times 7 + 6 = 34$	$3 \times 9 + 6 = 33$

2. a. 7 in. × 3 in. = 21 in^2 b. 25 km × 20 km = 500 km^2 c. 2 ft × 9 1/2 ft = 19 ft^2

3. a. acute   b. right   c. obtuse   d. right   e. obtuse   f. acute   g. right   h right

4. a. Answers vary. Check students' answers. For example:
   b. Answers vary. The measurements in this picture are not to scale.
   c. Answers vary. The angles of the parallelogram above are 58°, 122°, 58°, and 122°.

5. a. 268   b. 277

6. a. Four boxes weighing 44 lb each. They weigh 16 lb more.
      $4 \times 44 = 176$ and $5 \times 32 = 160$.
   b. The teacher got to keep 11 balloons. $86 \div 25 = 3$ R11.
   c. Five liters would cost $6.95. $9.73 \div 7 = 1.39. 5 \times 1.39 = 6.95$

Review, p. 168

1. a. 1   b. 5 1/8   c. 7   d. 2/10   e. 1 2/4   f. 6 7/12

2. a. 7/10   b. 3/5   c. 4/5   d. 5/8

3. a. 33/100   b. 53/100   c. 1 17/100

4. 1 3/4 liters

5. Answers vary. For example:

6.

a. ×2 $\frac{3}{4} = \frac{6}{8}$ ×2	b. ×5 $\frac{1}{2} = \frac{5}{10}$ ×5	c. $\frac{2}{5} = \frac{4}{10}$	d. $\frac{2}{3} = \frac{6}{9}$
		e. $\frac{2}{3} = \frac{8}{12}$	f. $\frac{3}{4} = \frac{12}{16}$

7. a. >   b. <   c. =   d. >   e. >   f. >   g. >   h. <

8. a. 9/10   b. 1 1/5   c. 1 4/10   d. 99/100   e. 2 4/8   f. 2 9/12

9.
Mexican Coffee (4x)
6 cups strong gourmet coffee
3 tsp cinnamon
16 tsp chocolate syrup
1 tsp nutmeg
2 cup heavy cream
4 tbsp sugar

10. a. 40; 80   b. 8 cm; 40 cm   c. 400 kg; $40

11. He has 1/4 of it left, which is $5.

12. 150 pages. One-eighth of 240 pages is 30 pages. She has 5/8 of the book left to read, which is $5 \times 30 = 150$ pages.

Decimals

Decimal Numbers—Tenths, p. 172

1. a. 0.7   b. 2.4   c. 10.9   d. 9/10   e. 29 3/10

2. a. 0.6 = 6/10   b. 1.2 = 1 2/10   c. 2.9 = 2 9/10

3.

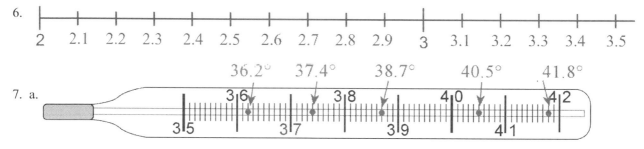

a. 0.4	c. 1.6
b. 0.1	d. 2.8

4. 7, 7 1/10, 7 2/10, 7 3/10, 7 4/10, 7 5/10, 7 6/10, 7 7/10, 7 8/10, 7 9/10, 8

5. 14.1, 14.2, 14.3, 14.4, 14.5, 14.6, 14.7, 14.8, 14.9

6.

7. a.

   b. The temperatures 38.7°, 40.5°, and 41.8° are fever.

8. a. <   b. >   c. >   d. <   e. =

9. 0.1 $\frac{1}{2}$ 0.9   1.2   2.3 $2\frac{1}{2}$ 2.6   3.0

Adding and Subtracting with Tenths, p. 174

1. a. 0.7 + 0.5 = 1.2   b. 0.6 + 0.8 = 1.4   c. 1.1 − 0.8 = 0.3   d. 1.3 − 0.4 = 0.9   e. 0.2 + 1.1 = 1.3

2. a. 9/10; 0.2 + 0.7 = 0.9   b. 1 1/10; 0.5 + 0.6 = 1.1   c. 1 7/10; 0.9 + 0.8 = 1.7

3. a. 1.1; 2.1   b. 1.2; 4.2   c. 1.5; 3.5   d. 0.9; 4.9

4. a. 3.2   b. 2.2   c. 6.1   d. 3

5. a. 5.3   b. 78.4   c. 63.4   d. 1.9, 8.9, 9.0 = 9, 9.1, 9.9

Adding and Subtracting with Tenths, continued

6.

a. 0.1	b. 1.1	c. 2.5	d. 3.6
+ 0.2 = 0.3	+ 0.5 = 1.6	+ 0.3 = 2.8	− 0.4 = 3.2
+ 0.2 = 0.5	+ 0.5 = 2.1	+ 0.3 = 3.1	− 0.4 = 2.8
+ 0.2 = 0.7	+ 0.5 = 2.6	+ 0.3 = 3.4	− 0.4 = 2.4
+ 0.2 = 0.9	+ 0.5 = 3.1	+ 0.3 = 3.7	− 0.4 = 2
+ 0.2 = 1.1	+ 0.5 = 3.6	+ 0.3 = 4.0	− 0.4 = 1.6
+ 0.2 = 1.3	+ 0.5 = 4.1	+ 0.3 = 4.3	− 0.4 = 1.2

7. a. Draw a line that is 4.7 cm long.   b. 2.4 cm

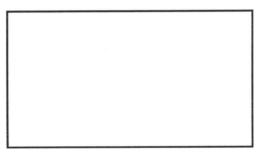

8. a. 5 mm; 12 mm   b. 0.7 cm; 3.5 cm   c. 1.4 cm; 7.4 cm

9. The perimeter is 20.2 cm.

Two Decimal Digits—Hundredths, p. 176

1. a. 0.08 = 8/100   b. 0.55 = 55/100   c. 1.50 = 1 50/100   d. 1.06 = 1 6/100   e. 3.70 = 3 70/100

2.

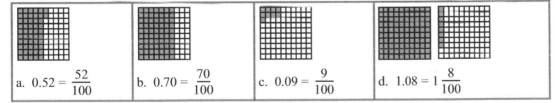

a. $0.52 = \frac{52}{100}$ b. $0.70 = \frac{70}{100}$ c. $0.09 = \frac{9}{100}$ d. $1.08 = 1\frac{8}{100}$

Teaching box:

> Now, **YOU DRAW** nine tiny lines between 0.2 and 0.3, dividing that distance into TEN new parts.
>
> If this process was repeated between 0.3 and 0.4, between 0.4 and 0.5, and so on,
> into how many parts would the number line from 0 to 1 be divided? _100_ parts
>
> These new parts are therefore **hundredth parts**, or **hundredths**.

3.
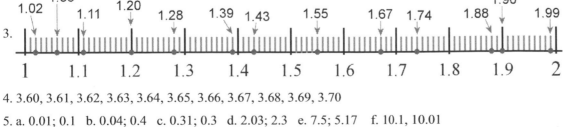

4. 3.60, 3.61, 3.62, 3.63, 3.64, 3.65, 3.66, 3.67, 3.68, 3.69, 3.70

5. a. 0.01; 0.1   b. 0.04; 0.4   c. 0.31; 0.3   d. 2.03; 2.3   e. 7.5; 5.17   f. 10.1, 10.01

Two Decimal Digits—Hundredths, continued

6.

	fraction	read as ...
a. 0.02	2/100	two hundredths
b. 1.49	1 49/100	one and forty-nine hundredths
c. 5.5	5 5/10	five and five tenths
d. 3.08	3 8/100	three and eight hundredths

7.

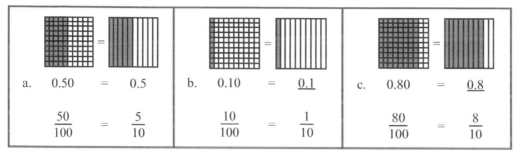

a. 0.50 = 0.5   b. 0.10 = 0.1   c. 0.80 = 0.8

$\frac{50}{100} = \frac{5}{10}$ $\frac{10}{100} = \frac{1}{10}$ $\frac{80}{100} = \frac{8}{10}$

8. $0.60

9. a. >   b. <   c. <   d. =

10. a. 7.9   b. 15.4 and 15.40 (they are equal)   c. 2.77   d. 9.3   e. 3.6   f. 0.4

11. a. >   b. =   c. >   d. =   e. <   f. >   g. <   h. >   i. >   j. <   k. >   l. <

12. a. 5.06 < 5.16 < 5.6 < 5.66   b. 7.70 < 7.77 < 7.78 < 7.8

Adding and Subtracting Hundredths, p. 180

1.

a. 0.05 + 0.04 = 0.09	b. 0.07 + 0.04 = 0.11	c. 0.37 − 0.06 = 0.31
$\frac{5}{100} + \frac{4}{100} = \frac{9}{100}$	$\frac{7}{100} + \frac{4}{100} = \frac{11}{100}$	$\frac{37}{100} - \frac{6}{100} = \frac{31}{100}$

d. 0.45 + 0.65 = 1.10	e. 3.25 − 1.08 = 2.17
$\frac{45}{100} + \frac{65}{100} = 1\frac{10}{100}$	$3\frac{25}{100} - 1\frac{8}{100} = 2\frac{17}{100}$

2. a. 0.12; 4.12   b. 0.95; 2.80   c. 1.0; 2.02   d. 0.02, 19.03

3.

a. 0.91	b. 0.80	c. 2.90	d. 1.77
+ 0.02 = 0.93	− 0.05 = 0.75	+ 0.03 = 2.93	+ 0.11 = 1.88
+ 0.02 = 0.95	− 0.05 = 0.70	+ 0.03 = 2.96	+ 0.11 = 1.99
+ 0.02 = 0.97	− 0.05 = 0.65	+ 0.03 = 2.99	+ 0.11 = 2.10
+ 0.02 = 0.99	− 0.05 = 0.60	+ 0.03 = 3.02	+ 0.11 = 2.21
+ 0.02 = 1.01	− 0.05 = 0.55	+ 0.03 = 3.05	+ 0.11 = 2.32
+ 0.02 = 1.03	− 0.05 = 0.50	+ 0.03 = 3.08	+ 0.11 = 2.43

4.

a. $0.97 + 0.04 = 1.01$ $0.95 + 0.11 = 1.06$	b. $2.96 + 0.06 = 3.02$ $8.91 + 0.11 = 9.02$	c. $1.03 - 0.04 = 0.99$ $1.12 - 0.16 = 0.96$	d. $7.02 - 0.05 = 6.97$ $4.01 - 0.50 = 3.51$

5. a. 3 cm   b. 45 cm   c. 109 cm   d. 282 cm   e. 9.80 m   f. 3.06 m

6. a. 1.65 m or 1 m 65 cm   b. 0.34 m or 34 cm taller   c. 6.60 m or 660 cm

Teaching box:

$0.2 + 0.05 = \underline{0.25}$

```
          0.2                                    0.3
  |    |    |    |    |    |    |    |    |    |    |    |    |    |    |
 0.18 0.19 0.20 0.21 0.22 0.23 0.24 0.25 0.26 0.27 0.28 0.29 0.30 0.31 0.32
```

If you are at 0.2 and go five hundredths (0.05) further, where will you end up?

7.

a. $0.7 \;+\; 0.04$ $\downarrow \qquad \downarrow$ $0.70 + 0.04 = 0.74$	b. $0.5 \;+\; 0.11$ $\downarrow \qquad \downarrow$ $0.50 + 0.11 = 0.61$

8.

a. $0.1\underline{0} + 0.05 = \underline{0.15}$ $\dfrac{10}{100} + \dfrac{5}{100} = \dfrac{15}{100}$	b. $0.04 + 0.4\underline{0} = 0.44$ $\dfrac{4}{100} + \dfrac{40}{100} = \dfrac{44}{100}$	c. $0.6\underline{0} - 0.09 = 0.51$ $\dfrac{60}{100} - \dfrac{9}{100} = \dfrac{51}{100}$
d. $0.6\underline{0} + 0.22 - \underline{0.82}$ $\dfrac{60}{100} + \dfrac{22}{100} = \dfrac{82}{100}$	c. $0.73 - 0.5\underline{0} = 0.23$ $\dfrac{73}{100} - \dfrac{50}{100} = \dfrac{23}{100}$	f. $0.9\underline{0} - 0.13 - 0.77$ $\dfrac{90}{100} - \dfrac{13}{100} = \dfrac{77}{100}$

9.

a. $0.11 + 0.5\underline{0} = 0.61$	b. $0.24 - 0.2\underline{0} = 0.04$	c. $0.3\underline{0} + 0.39 = 0.69$
d. $0.22 + 0.7\underline{0} = 0.92$	e. $0.6\underline{0} - 0.41 = 0.19$	f. $0.97 - 0.7\underline{0} = 0.27$

10. a. 0.4; 0.33   b. 1.4; 1.43   c. 3.77; 3.87

11. a. 100 ml; 700 ml; 400 ml   b. 700 ml; 0.7 L

12.

1.6	1.21	1.3	1.3	1.18	1.45	1.45
1.7	1.23	1.24	0.7	1.24	1.25	1.51
1.52	1.03	1.18	1.38	1.59	1.31	1.71
1.43	0.94	1.02	1.32	1.92	1.72	1.66
2.2	1.95	1.72	1.52	1.5	1.7	1.9
1.8	1.6	1.66	1.48	1.3	1.64	1.58
1.74	1.4	1.78	1.71	1.28	1.98	1.78

Puzzle corner. a. $x = 0.15$ b. $x = 0.06$ c. $x = 0.18$

Adding and Subtracting Decimals in Columns, p. 184

1. a. 62.29   b. 19.28   c. 183.39

2. a. 13.99   b. 49.89   c. 12.16

3. a. >   b. <   c. <   d. <   e. >   f. <   g. <   h. >

4. a. 14.03   b. 70.64   c. 4.84

5. a. 0.82   b. 15.63

6.

a. Mary did not line up the decimal points correctly. The correct answer: $\begin{array}{r} 4\ 5\ .\ 5 \\ +\ \ 5\ .\ 3\ 4 \\ \hline 5\ 0\ .\ 8\ 4 \end{array}$	b. Jack did not regroup at all. The correct answer: $\begin{array}{r} 9 \quad\ \ 9 \\ 8\ \cancel{10}\ \ \cancel{10}\ 10 \\ \cancel{9}\ \cancel{0}\ .\ \cancel{0}\ \cancel{0} \\ -\ 8\ 8\ .\ 5\ 6 \\ \hline 1\ .\ 4\ 4 \end{array}$

7. 2.78 kg

8. 11.25 m

9. 2.65 kg

Puzzler corner   a. 4.8 + 40.8 + 4.08 = 49.68   b. 560 − 5.06 − 56 = 498.94

Using Decimals with Measuring Units, p. 187

1. a. 700 ml; 0.7 L   b. 300 ml; 0.3 L   c. 200 ml; 500 ml; 5,400 ml   d. 0.1 L; 1.5 L; 6.3 L

2. a. 0.4 km or 400 m   b. 600 m: 1,100 m   c. 0.7 km; 1.8 km   d. 10,900 m; 24.6 km

3. a. 0.6 kg; 2.4 kg   b. 200 g; 800 g   c. 20.5 kg; 7,100 g

4. a. 350 g   b. 2.25 kg or 2,250 g

5. Julie walks 1.8 km more. (Julie walks 2 × 1.2 km = 2.4 km. Amanda walks 2 × 0.3 km = 0.6 km.)

6. 7.3 liters, or 7,300 ml

7. The perimeter is 450 cm. One side is 90 cm.

8. 1.3°F; 1.8°F

Mixed Review, p. 189

1. a. 5/8 + 3/8 + 2/8 = 1 2/8   b. 1 7/12 + 7/12 = 2 2/12

2. a. 50/12 = 4 2/12   b. 28/9 = 3 1/9   c. 42/100

3. a. 1, 2, 19, 38   b. 1, 2, 4, 7, 8, 14, 28, 56   c. 1, 19 (it is prime)

4. a. ≈ 6 × 300 = 1,800 Exact: 1,752   b. ≈ 11 × 400 = 4,400   Exact 4,422
   c. ≈ 3 × 2,400 = 7,200   Exact 7092   d. ≈ 7 × 9,000 = 63,000   Exact 61,789

5. a. 90   b. 4   c. 30

6. a. 2:30 pm   b. 7:15 pm   c. 10:45 pm   d. 7:50 am

Mixed Review, continued

7.

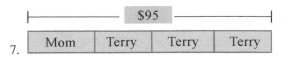

Mom paid $23.75 and Terry paid $71.25.

8. He still has to pay $240. First find 1/5 of $600, which is $120. Jack still has to pay 2/5 of the price, which is 2 × $120 = $240.

9.

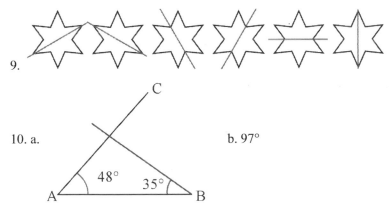

10. a.

b. 97°

Review, p. 191

1. a. 0.7   b. 0.07   c. 1.6   d. 2.41   e. 1.01   f. 0.47   g. 8/10   h. 2 9/10   i. 4 14/100   j. 18 8/100   k. 3/100   l. 29/100

2. a. >   b. >   c. <   d. =   e. >   f. =   g. <   h. <   i. >

3.

```
0.02  0.09              0.38    0.47    0.60                    0.91
 |  |                    |       |       |                      |
0     0.1   0.2   0.3   0.4   0.5   0.6   0.7   0.8   0.9   1
```

4. 0.1, 0.12, 0.2, 0.21, 1/2, 0.74, 0.8

5.

a. 0.70 + 0.03 = 0.73	b. 0.32 + 0.40 = 0.72	c. 0.70 − 0.04 = 0.66
$\frac{70}{100} + \frac{3}{100} = \frac{73}{100}$	$\frac{32}{100} + \frac{40}{100} = \frac{72}{100}$	$\frac{70}{100} - \frac{4}{100} = \frac{66}{100}$

6. a. 1   b. 0.88   c. 0.36   d. 0.24   e. 0.83   f. 0.5

7. a. incorrect. Should be: 0.99 + 0.1 = 1.09   OR   0.99 + 0.01 = 1
   b. correct
   c. incorrect. Should be: 0.19 + 0.19 = 0.38
   d. incorrect. Should be: 0.03 + 0.5 = 0.53   OR   0.03 + 0.05 = 0.08

8. a. 9.31   b. 23.11   c. 5.84

9. 2.84

10. 900 grams; 200 grams; a tablet weighing 610 grams is heavier

Test Answer Keys

Math Mammoth Grade 4 Tests Answer Key

Chapter 1 Test

1. $x = 2,611$

2. a. 260   b. 20   c. 70

3. The expression $\$20 - 7 \times \$2$ matches the problem. The answer is $6.

4. $\$30 + 2 \times \$14 = \$58$ — $50,00 estimate

5. $10\ m + 8\ m + 15\ m = 33\ m$

6. $\$67 + \$48 = x$; $x = \$115$.
$\longleftarrow$ original price $\boxed{x}$ $\longrightarrow$
| $67 | $48 |

Chapter 2 Test

1. a. 400,040   b. 64,500   c. 200,067

2. a. eighty thousand or 80,000   b. eighty or 80

3. a. 516,800   b. 293,000   c. 200,000

4. 207,698

5. 3,294   39,244   39,294   93,294   399,295

6. The king of Nootyland has more coins; he has 5,218 coins more. The king of Sookiland has $3 \times 24,000 + 1,382$ = 73,382 coins. The king of Nootyland has 78,600 coins; the difference is $78,600 - 73,382 = 5,218$.

7. A thousand students.

Chapter 3 Test

1. a. $40 + 32 = 72$ b. $140 + 42 = 182$ c. $2,100 + 27 = 2,127$

2. About $7 \times \$20 = \$140$.

3. a. 6,300   b. 6,000   c. 160,000

4. a. 200   b. 200   c. 8

5. a. 18,000   b. 48,000   c. 1,293   d. 2,080

6. a. 1,170   b. 5,848   c. 1,045   d. 15,924

7. 1,360

8. a. One meal costs $3, so seven meals cost $21.   b. $\$30 - 7 \times \$2.55 = \$30 - \$17.85 = \$12.15$
   c. She took $3 \times \$12.55 + \$8.90 + \$13.45 = \$60$ d. It would have cost $30 ($\$150 \div 5 = \$30$).

Chapter 4 Test

1. 3:50 p.m.

2. a. 2 3/8 in. or 6 cm 0 mm b. 3 7/8 in. or 9 cm 8 mm

3.

a.	b.	c.
4 lb 2 oz = 66 oz 76 cm = 760 mm 5 ft 5 in = 65 in.	2 L 80 ml = 2,080 ml 3 qt = 12 cups 200 yd = 600 ft	7 m 5 cm = 705 cm 4 kg 500 g = 4,500 g 3 T = 6,000 lb

4. 10 cm 4 mm

5. a. two bottles   b. five bottles.

6. a. $7.60   b. $10.50

7. eight jars

Chapter 5 Test

1. a. 3 R1; 2 R3   b. 4 R5; 5R5   c. 3 R4; 7R4

2. One meter costs $6, so five meters would cost 5 × $6 = $30.

3. $210. One-fifth of $350 is $70; three-fifths of that is 3 × $70 = $210.

4. 250 bricks. One-third of his 1,200 bricks is 400 bricks, and two-thirds of them is double that, or 800 bricks. So, he has 400 bricks left. After selling 150 bricks, he has 250 bricks left.

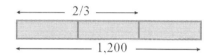

5. a. 113. Check: 5 × 113 = 565   b. 458   Check: 8 × 458 = 3,664

6. a. 1, 2, 4, 7, 14, 28   b. 1, 13   c. 1, 2, 4, 8, 16, 32   d. 1, 2, 4, 19, 38, 76

7. 125 ÷ 7 = 17 R6. Each child got 17 pencils, and 6 pencils were left over.

8. $47. Add the prices and divide by four: average = ($39 + $45 + $63 + $41) ÷ 4 = $47.

9. Yes, it is, because 924 ÷ 7 = 132 and there is no remainder; the division is even.

10. (30 − 10) × 20 = 400

Chapter 6 Test

1. Check students' answers. Here is a 75° angle:

2. a. 140°

3. It is 148 degrees.

   $78° + 134° + x = 360°$

   $x = 360° - 134° - 78°$

   $= 148°$

4. a. a parallelogram or a rhombus (either is correct)

b.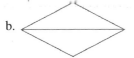

   c. 17 cm 7 mm

5. Answers vary. Check students' answers. For example:

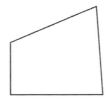

6. Answers vary. Check students' answers. The two other angles should have an angle sum of 90°.

7. a. a right triangle   b. an acute triangle

8. Subtract the area of the outer rectangle and the area of the white rectangle: 20 ft × 10 ft − 10 ft × 3 ft
   = 200 ft² − 30 ft² = 170 ft².

Chapter 7 Test

1. a. 1   b. 2 1/3   c. 5 4/5   d. 2/12   e. 2 3/5   f. 5 4/6

2. a. $\dfrac{3}{8}, \dfrac{1}{2}, \dfrac{3}{4}$ b. $\dfrac{5}{7}, \dfrac{5}{5}, \dfrac{7}{5}$ c. $\dfrac{5}{9}, \dfrac{5}{6}, \dfrac{5}{2}$

3.

a. Split all pieces into four new ones	b. Split all pieces into three new ones.
$\dfrac{1}{2} = \dfrac{4}{8}$	$\dfrac{2}{3} = \dfrac{6}{9}$

4.

a. $\dfrac{1}{5} = \dfrac{2}{10}$	b. $\dfrac{3}{4} = \dfrac{9}{12}$	c. $\dfrac{4}{5} = \dfrac{20}{25}$	d. $\dfrac{1}{6} = \dfrac{4}{24}$

5. a. 12/10 = 1 2/10   b. 15/5 = 3   c. 12/8 = 1 4/8 (which is equal to 1 1/2)

6. a. Walter and Eric ate equal amounts. Walter ate 1/4 of it, which is equal to 3/12, or 3 pieces.
   b. Eric ate 3/12 and John ate 1/12. So, Eric ate 2/12 of the pizza more than John

1.

A number line from 1 to 2 with arrows pointing to: 1.04, 1.21, 1.60, 1.78

2. a. 0.2   b. 7.04   c. 0.74   d. 52/100   e. 3 9/10

3. a. 2.2   b. 0.95   c. 0.19   d. 0.7   e. 0.37   f. 3.04

4. a. >   b. =   c. >   d. <   e. <

5. $2.07 < 2.17 < 2.7 < 2.77 < 7.2$

6. 5.2 kg. You can add 1.3 kg + 1.3 kg + 1.3 kg + 1.3 kg = 5.2 kg

7. a. 7.36   b. 1.76

1. 1,980. Add to check: 1,980 + 543 + 2,677 equals 5,200.

2. a. ≈ $1 + $9 + $4 + $9 = $23
   b. Her bill is $1.28 + $8.92 + $3.77 + $9.34 = $23.31. Her change is $30 − $23.31 = $6.69.

3. Estimate: 5 × $0.90 + 2 × $1.20 = $4.50 + $2.40 = $6.90

4. a. 30; 84   b. 11; 14   c. 140; 19

5. a. $35 + x = $92 ; x = $57   b. x − 24 = 37 ; x = 61

6. a. 2,000   1,750   1,500   1,250   1,000   750   500   250

   b. 200, 500, 800, 1100, 1400, 1700

7. In the frequency table we list how many students got that score.

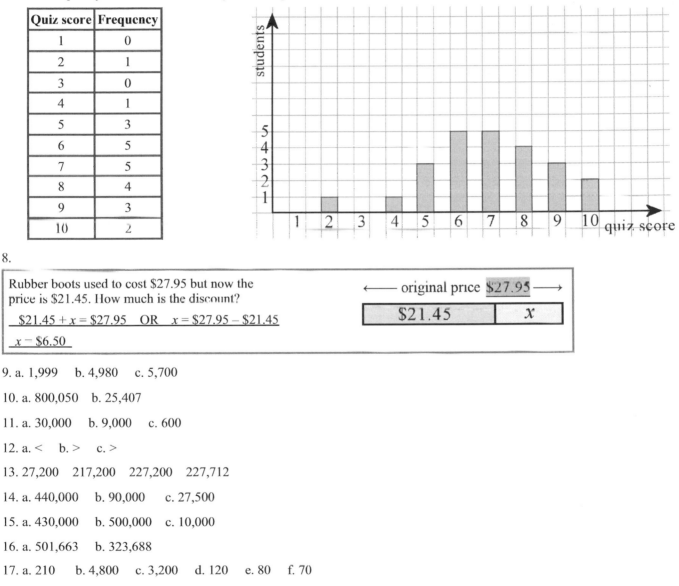

Quiz score	Frequency
1	0
2	1
3	0
4	1
5	3
6	5
7	5
8	4
9	3
10	2

8.

Rubber boots used to cost $27.95 but now the price is $21.45. How much is the discount?

$21.45 + x = $27.95   OR   x = $27.95 − $21.45

x − $6.50

←——— original price $27.95 ———→

$21.45	x

9. a. 1,999   b. 4,980   c. 5,700

10. a. 800,050   b. 25,407

11. a. 30,000   b. 9,000   c. 600

12. a. <   b. >   c. >

13. 27,200   217,200   227,200   227,712

14. a. 440,000   b. 90,000   c. 27,500

15. a. 430,000   b. 500,000   c. 10,000

16. a. 501,663   b. 323,688

17. a. 210   b. 4,800   c. 3,200   d. 120   e. 80   f. 70

18. a. $160   b. $800   c. four days, since 4 × $160 = $640

19. a. estimate 5 × 200 = 1,000. Exact: 980
    b. estimate 40 × 40 = 1,600 or 30 × 40 = 1,200. Exact: 1,330
    c. estimate 7 × 3,000 = 21,000. Exact: 22,316
    d. estimate 90 × 20 = 1,800. Exact: 1,958

20.

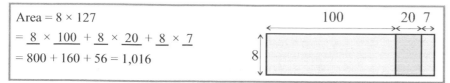

Area = 8 × 127
= 8 × 100 + 8 × 20 + 8 × 7
= 800 + 160 + 56 = 1,016

21. a. Answers may vary. For example: $400 − 26 × $14 = $400 − $364 = $36. Or, 26 × $14 = $364 and $400 − $364 = $36.
    b. 24 × 60 minutes = 1,440 minutes
    c. Answers may vary. For example: 4 × 375 cm = 1,500 cm. Or, 375 cm + 375 cm + 375 cm + 375 cm = 1,500 cm
    d. Answers may vary. For example: ($277 − $58) × 8 = $1,752. Or, $277 − $58 = $219 and 8 × $219 = $1,752.

22. Answers may vary if the test is printed with "shrink to fit" or "fit to printable area", or because of slight variability in
    rulers, or because of measuring inaccurately. Please check students' answers.
    a. 5 1/4 in. or 13 cm 3 mm. 13 cm 4 mm is also acceptable.   b. 3 7/8 in. or 9 cm 8 mm. 9 cm 9 mm is also acceptable.

23. 6 hours 12 minutes

24. 1 h 45 min + 50 min + 1 h 15 min + 2 h 15 min + 55 min = 4 h 180 min, which is 7 hours.

25. She worked 7 hours 30 minutes. From 7:00 am till 3:35 pm is 8 hours 35 minutes. Subtract from that 65 minutes,
    or 1 hour 5 minutes, to get 7 hours 30 minutes.

26.

a.	b.	c.
6 lb = 96 oz 2 lb 11 oz = 43 oz	5 gal = 20 qt 2 qt = 8 cups	4 ft 2 in. = 50 in. 7 yd = 21 ft

27.

a.	b.	c.
2 kg = 2,000 g 11 kg 600 g = 11,600 g	5 L 200 ml = 5,200 ml 3 m = 300 cm	8 cm 2 mm = 82 mm 10 km = 10,000 m

28. In four days, he jogs 15 km 200 m.

29. 1 L 650 ml

30. 17 ft 8 in

31. a. 63. Check: 63 × 9 = 567   b. 2,141. Check: 2141 × 4 = 8,564

32. a. 9 R2   b. 8 R1   c. 6 R3

33. a. Three photos on the last page; five pages were full.
    b. Your neighbor should be $36, because one foot of the fence costs $3.

34. a. It cost $99. First find 1/8 of $264: $264 ÷ 8 = $33. Then to find 3/8 of it, multiply 3 × $33 = $99.
    b. She needs 20 bags. 117 ÷ 6 = 19 R3. Notice she needs a bag also for the three muffins that don't fill a bag.

35.

number	divisible by 1	divisible by 2	divisible by 3	divisible by 4	divisible by 5	divisible by 6	divisible by 7	divisible by 8	divisible by 9	divisible by 10
80	x	x		x	x			x		x
75	x		x		x					
47	x									

36.

a. Is 5 a factor of 60?	b. Is 7 a divisor of 43?
Yes , because 5 × 12 = 60 .	No , because 43 ÷ 7 = 6 R1 (the division is not even).
c. Is 96 divisible by 4? Yes , because 96 ÷ 4 = 24 (the division is even).	d. Is 34 a multiple of 7? No , because 34 is not in the multiplication table of 7. OR: No, because 34 ÷ 7 = 4 R6; the division is not even. OR: No, because there is no whole number you can multiply by 7 to get 34.

37. Answers vary. For example: 2, 3, and 5. Here is a list of primes less than 100:
  2 3 5 7 11 13 17 19 23 29 31 37 41 43 47 53 59 61 67 71 73 79 83 89 97

38. a. 1, 2, 4, 7, 8, 14, 28, 56   b. 1, 2, 3, 6, 13, 26, 39, 78

39. 155°

40. Check students' answers.

41. Answers vary. Check students' answers. The sum of the angle measures should be 180° or very close.

42. 29° + x = 180°;   x = 151°.

43. Right angles.

44. Answers vary. Check students' answers. For example:

45.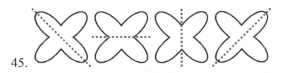

46. Use subtraction. A = 28 ft × 12 ft − 6 ft × 10 ft = 336 ft^2 − 60 ft^2 = 276 ft^2.

47. $\dfrac{5}{8}$ + $\dfrac{5}{8}$ = $1\dfrac{2}{8}$

48. There are still 2/4 or 1/2 of it left to do.

49. a. 1 2/5   b. 5/6   c. 6

50.

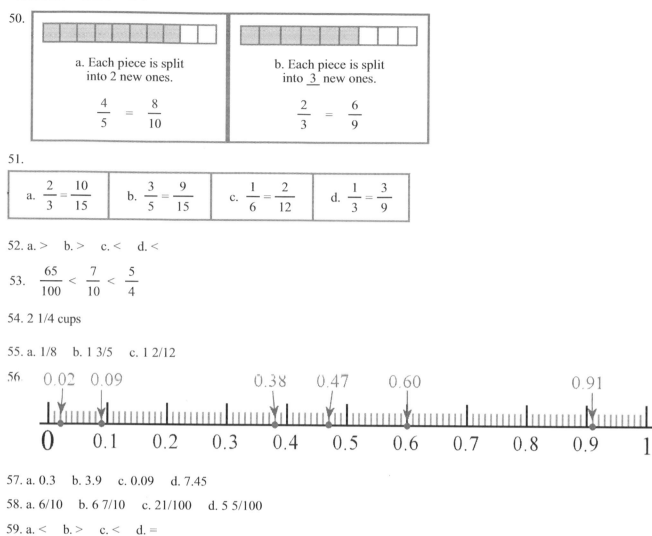

a. Each piece is split into 2 new ones.

$$\frac{4}{5} = \frac{8}{10}$$

b. Each piece is split into _3_ new ones.

$$\frac{2}{3} = \frac{6}{9}$$

51.

a. $\frac{2}{3} = \frac{10}{15}$ b. $\frac{3}{5} = \frac{9}{15}$ c. $\frac{1}{6} = \frac{2}{12}$ d. $\frac{1}{3} = \frac{3}{9}$

52. a. >   b. >   c. <   d. <

53. $\frac{65}{100} < \frac{7}{10} < \frac{5}{4}$

54. 2 1/4 cups

55. a. 1/8   b. 1 3/5   c. 1 2/12

56.

0.02 0.09   0.38   0.47   0.60   0.91

0   0.1   0.2   0.3   0.4   0.5   0.6   0.7   0.8   0.9   1

57. a. 0.3   b. 3.9   c. 0.09   d. 7.45

58. a. 6/10   b. 6 7/10   c. 21/100   d. 5 5/100

59. a. <   b. >   c. <   d. =

60. a. 13.01   b. 3.74

Cumulative Reviews
Answer Keys

Cumulative Reviews Answer Key, Grade 4

Cumulative Review: Chapters 1 - 2

1. a. 138; 74; 103   b. 58; 92; 144   c. 127; 70; 144

2. a. 1,800; 1,000   b. 600; 510   c. 9; 1   d. 500; 140

3. a. Continue this pattern: subtract _80_ each time.

| 700 | 620 | 540 | 460 | 380 | 300 | 220 | 140 |

b.

| 0 | 99 | 198 | 297 | 396 | 495 | 594 | 693 |

4. Addition: $450 + 128 + x = 1,000$; Solution $x = 1,000 - 128 - 450 = 422$

1,000		
450	128	x

5. a. 4,445   b. 13,378   c. 716,051

6. a. $8,030 < 18,399 < 818,939 < 819,090$

   b. $5,220 < 52,200 < 250,500 < 520,500$

7. a. 284 thousand 1   b. 50 thousand 50

| 2 | 8 | 4, | 0 | 0 | 1 |

| | 5 | 0, | 0 | 5 | 0 |

8. a. 2,000   b. 10   c. 500,000   d. 40,000

9. a. 7,500   b. 2,700   c. 4,000   d. 400   e. 56,300   f. 293,600

10. $176 + $25 + $30 = $231.

Cumulative Review: Chapters 1 - 3

1. a. 9; 59; 590; 190   b. 6; 66; 160; 600   c. 5; 75; 500; 550

2. $458 + $366 + $427 + $503 + $413 = $2,167

3. a.

Kilometers	80	400	560	720	800	960	1,200	1,600
Hours	1	5	7	9	10	12	15	20

b.

Dollars	$9	$18	$27	$36	$45	$72	$90	$135
Yards	1	2	3	4	5	8	10	15

4. a. $1,554 < 5,000 < 5,005 < 5,500 < 5,604$

   b. $3,800 < 37,700 < 38,707 < 73,737 < 307,988$

5. a. 983,177   b. 555,330

6. a. $3.05 \approx $3.00   b. $8.32 \approx $8.00   c. $25.97 \approx $26.00

7. a. Approximately $130 + $75 + $90 + $140 + $70 = $505.
   b. They earned about $65 less on their worst day than on their best day. (Worst day: about $75, best day about $140.)

8. a. He read $12 \times 96 = 1,152$ pages.
   b. Half of the magazines is 6 magazines. Jesse read six magazines in $6 \times 2\ 1/2$ hours = 15 hours.

1. a. 975   b. 20,990   c. 1,968   d. 4,088

2. $256 + x = 609$; $x = 353$.

3. a. 27   b. 57   c. 22   d. 57   e. 21   f. 1

4. a. $555 \approx 600$ b. $8,889 \approx 8,900$ c. $351,931 \approx 351,900$
    d. $64 \approx 100$ e. $244,295 \approx 244,300$ f. $38,009 \approx 38,000$

5. a. 305,200   b. 40,033

6.

a. $723,050 > 699,099$	b. $322,320 < 322,322$
c. $692,159 < 692,196$	d. $140,000 > 14,100$
e. $113,999 < 115,399$	f. $836,496 > 88,482$

7. a. 1,100; 190; 120,000   b. 180; 800,000; 8,800   c. 92,000; 64,000; 8,800

8.

| a. $6 \times 30¢$ $= 180¢ = \$1.80$ | b. $5 \times 84¢$ $= 400¢ + 20¢ = \$4.20$ |
| c. $6 \times \$1.70$ $= \$6.00 + \$4.20 = \$10.20$ | d. $3 \times \$2.80$ $= \$6.00 + \$2.40 = \$8.40$ |

9. a. about $20 \times 40 = 800$ plants
   b. The cost was $7 \times \$8.20 = \$56 + \$1.40 = \$57.40$. His change was $\$100 - \$57.40 = \$42.60$.

1. a. 1,000 meters are not accessible by boat. (Each little "block" in the diagram is 200 m.)

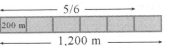

b. There are 22 girls. There are 33 boys and girls.

2. a. The baker spent about $90 more for flour in May than in March. In May, he spent about $550 and in March, about $460.

   b. Estimates may vary. He spent about $460 + $620 + $550 = $1,630.

3. a. One triangle weighs 3 units. Solution: First take off two triangles from both sides. That leaves: 11 = 8 + triangle. So, one triangle has to equal 3.

   b. One square weighs 2 units. Solution: Take off one square from both sides. That leaves: 3 squares + 9 = 15. Now, take off "9" from both sides. That leaves 3 squares = 6. So, one square has to weigh 2.

4. a. 1 kg 300 g = 1,300 g   b. 3 lb = 48 oz   c. 7,500 g = 7 kg 500 g
       4 kg 20 g = 4,020 g   7 lb = 112 oz   4 lb 8 oz = 72 oz

5.

Minutes	1	5	6	7	10
Seconds	60	300	360	420	600

Days	1	3	6	10
Hours	24	72	144	240

6. First do 5 + 39 , which equals 44 . Then, divide that answer by 4 .

   This leaves 11 . Then, do 2 × 2 = 4 .

   Lastly, subtract that from 11 . The answer is 7 .

7. a. 4 h 54 m   b. 7 h 24 m   c. 8 h 24 m

8.

a.	b.	c.	d.
2 pt = 4 C	1 qt = 4 C	6 L = 6,000 ml	2 L 560 ml = 2,560 ml
2 C = 16 oz	2 gal = 8 qt	1/4 L = 250 ml	1,300 ml = 1 L 300 ml

9. a. January, February, March, April, September, October, November, and December (the temperature is below freezing, or below 32°).
   b. November, December, and January
   c. 37°

1. a. $18 + x = \$33$; $x = \$15$.   The unknown x is how much Dana earned.
   b. $100 - \$86 = x$; $x = \$14$. The unknown x is Dad's change.
   c. $120 - 39 = x$; $x = 81$.   The unknown x is the number of eggs that broke.
   d. $13 + 43 = x$; $x = 56$. The unknown x is the number of dogs the shelter had initially.

2. a. 590. Check: $590 \times 3 = 1,770$ b. 878. Check: $878 \times 9 = 7,902$.

3. a. 5 R3; 2 R9   b. 5 R3; 3 R8   c. 9 R2; 11 R1

4.

a. 7×78	b. 13×67	c. 311×8
$\approx 7 \times 80 = 560$	$\approx 13 \times 70 = 910$ OR $\approx 10 \times 70 = 700$	$\approx 300 \times 8 = 2,400$ OR $\approx 310 \times 8 = 2,480$

5. a. $60,000 + 70 = 60,070$ b. $123,000 + 4,000 + 4 = 127,004$
   c. $3 + 90,000 + 40 = 90,043$ d. $7 + 20 + 632,000 = 632,027$

6.

a. 7 m = 700 cm 69 mm = 6 cm 9 mm	b. 2 m 6 cm =206 cm 6 km = 6,000 m	c. 4 km 100 m = 4,100 m 169 cm = 1 m 69 cm

7.

a. 3 lb 8 oz = 56 oz 4 kg 11 g = 4,011 g	b. 32 oz = 2 lb 4,900 g = 4 kg 900 g	c. 7 lb 2 oz = 114 oz 36 kg 140 g = 36,140 g

8. a. $205   b. 2 km 600 m   c. She spent $7.12. Her change was $2.88

1.

a. 22,934 + 5,312 + 424,787	b. 519,313 − 47,616
Estimation: 23,000 + 5,000 + 420,000 = 448,000	Estimation: 520,000 − 50,000 = 470,000
Calculation: 453,033	Calculation: 471,697

2. 45 × 22 + 27 = 1,017 students

3. Right angles are exactly <u>90</u>°.
   Right triangles have exactly <u>one</u> right angle.

   Obtuse angles are more than <u>90</u>°, but less than <u>180</u>°.
   Obtuse triangles have exactly <u>one</u> obtuse angle.

   Acute angles are less than <u>90</u>°.
   Acute triangles have <u>three</u> acute angles.

4. Answers vary. To find possible side lengths, remember that the two side lengths add up to 14 in.

One side	Other side	Perimeter	Area
2 in.	12 in.	28 in.	24 sq. in.
3 in.	11 in.	28 in.	33 sq. in.
4 in.	10 in.	28 in.	40 sq. in.
5 in.	9 in.	28 in.	45 sq. in.

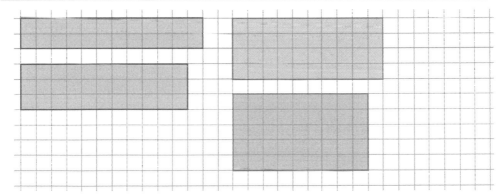

5.

a. 1:40 p.m.	b. 9:20 p.m.	c. 2:15 p.m.	d. 10:04 a.m.
13 : 40	21 : 20	14 : 15	10 : 04

6.

number	divisible by 2	by 5	by 10
478	x		
540	x	x	x
255		x	

number	divisible by 2	by 5	by 10
1,492	x		
3,093			
94	x		

number	divisible by 2	by 5	by 10
904	x		
905		x	
906	x		

7. No, because the division leaves a remainder: 549 ÷ 7 = 78 R3

8. a. 1, 2, 3, 4, 6, 8, 12, 24   b. 1, 2, 3, 6, 11, 22, 33, 66
   c. 1, 2, 3, 4, 6, 8, 12, 16, 24, 32, 48, 96   d. 1, 3, 5, 15, 25, 75

9.

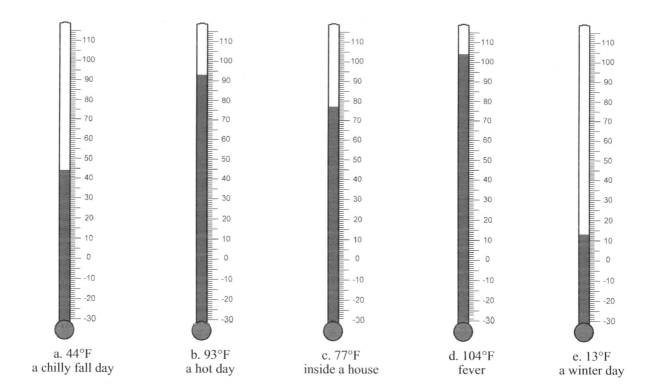

| a. 44°F | b. 93°F | c. 77°F | d. 104°F | e. 13°F |
| a chilly fall day | a hot day | inside a house | fever | a winter day |

1.

a. 4 × 36 120 + 24 = 144	b. 5 × 65 300 + 25 = 325	c. 8 × 426 3,200 + 160 + 48 = 3,408

2. 98,889

3. Answers vary since estimations can be done in various ways.

a. 8 × 69 ≈ 8 × 70 = 560	b. 11 × 55 ≈ 10 × 60 = 600 OR 11 × 60 = 660 OR 10 × 55 = 550	c. 25 × 17 ≈ 25 × 20 = 500 OR 24 × 20 = 480 (better, rounding one down, one up)

4. a. $x ÷ 8 = 7$; $x = 56$ b. $24 ÷ x = 8$; $x = 3$

5. yes - no - no - yes - no - yes - no

6.

a. 5 ft = 60 in. 12 ft = 144 in.	b. 3 ft 4 in. = 40 in. 6 ft 6 in. = 78 in.	c. 4 yd = 12 ft 9 yd = 27 ft

7. a. 5 kg 500 g   b. 3 kg 400 g   c. 9,900 g

8. a. Their total weight was 1 lb 1 oz. 3 oz + 3 oz + 5 oz + 2 oz + 4 oz = 17 oz = 1 lb 1 oz.

   b. 44 boxes. The division is 175 ÷ 4 = 43 R3. Keep in mind, she needs to pack the "leftover" 3 kg also into one box.

   c. She has $84 left. $98 ÷ 7 = 14; $98 − $14 = $84.

   d. Either 2, 3, 4, 6, 9, 12, or 18 children in a row. Probably 4, 6, and 9 children in a row are most practical.

   e. There are 34 foals, and 102 horses in total.

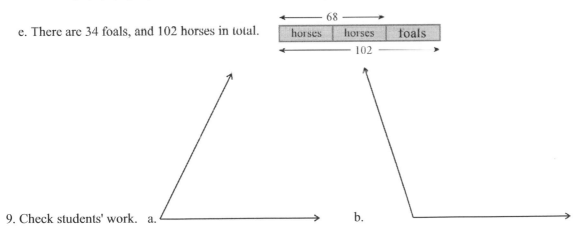

9. Check students' work. a.   b.

10. $o \parallel AB$, $o \parallel m$, $AB \parallel m$. There are no perpendicular lines, rays, or line segments.

11. Three-fourths of it are left.

12. a. 3   b. 1 1/12   c. 13/100   d. 1 1/4   e. 26/100   f. 6 1/10

13. a. 3   b. 1 1/5   c. 3 3/4   d. 7

14.

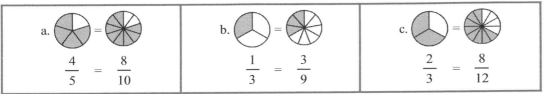

15. a. <   b. <   c. <   d. >

Made in the USA
Charleston, SC
23 June 2013